Text Book of

# MOBILE AND WIRELESS COMMUNICATION

**For**
**Third Year Diploma – Semester VI**
**Diploma in Electronics Engineering Groups**
**(ECE 605)**

***As Per New Syllabus of SBTE, Jharkhand***

**Vijay G. Yangalwar**
*M.Sc. LL.B., L.M.I.S.T.E.*
Retd. HOD, Deptt. of Electronics Engineering,
Government Polytechnic, Sadar, Nagpur.

N4748

**Mobile and Wireless Communication** **ISBN 978-93-89944-34-1**

**First Edition : February 2020**



**Published By :**
**NIRALI PRAKASHAN**
Abhyudaya Pragati, 1312, Shivaji Nagar
Off J.M. Road, PUNE – 411005
Tel - (020) 25512336/37/39, Fax - (020) 25511379
Email : niralipune@pragationline.com

## ➢ DISTRIBUTION CENTRES

**PUNE**

**Nirali Prakashan** (For orders within Pune) : 119, Budhwar Peth, Jogeshwari Mandir Lane, Pune 411002, Maharashtra
Tel : (020) 2445 2044, Fax : (020) 2445 1538; Mobile : 9657703145
Email : niralilocal@pragationline.com

**Nirali Prakashan** (For orders outside Pune) : S. No. 28/27, Dhayari, Near Asian College Pune 411041
Tel : (020) 24690204 Fax : (020) 24690316; Mobile : 9657703143
Email : bookorder@pragationline.com

**MUMBAI**

**Nirali Prakashan** : 385, S.V.P. Road, Rasdhara Co-op. Hsg. Society Ltd.,
Girgaum, Mumbai 400004, Maharashtra; Mobile : 9320129587
Tel : (022) 2385 6339 / 2386 9976, Fax : (022) 2386 9976
Email : niralimumbai@pragationline.com

## ➢ DISTRIBUTION BRANCHES

**JALGAON**

**Nirali Prakashan** : 34, V. V. Golani Market, Navi Peth, Jalgaon 425001, Maharashtra,
Tel : (0257) 222 0395, Mob : 94234 91860; Email : niralijalgaon@pragationline.com

**KOLHAPUR**

**Nirali Prakashan** : New Mahadvar Road, Kedar Plaza, 1st Floor Opp. IDBI Bank, Kolhapur 416 012
Maharashtra. Mob : 9850046155; Email : niralikolhapur@pragationline.com

**NAGPUR**

**Nirali Prakashan** : Above Maratha Mandir, Shop No. 3, First Floor,
Rani Jhanshi Square, Sitabuldi, Nagpur 440012, Maharashtra
Tel : (0712) 254 7129; Email : pratibhabookdistributors@gmail.com

**DELHI**

**Nirali Prakashan** : 4593/15, Basement, Agarwal Lane, Ansari Road, Daryaganj
Near Times of India Building, New Delhi 110002 Mob : 08505972553
Email : niralidelhi@pragationline.com

**BENGALURU**

**Nirali Prakashan** : Maitri Ground Floor, Jaya Apartments, No. 99, 6th Cross, 6th Main,
Malleswaram, Bengaluru 560003, Karnataka; Mob : 9449043034
Email: niralibangalore@pragationline.com

**Other Branches : Hyderabad, Chennai**



niralipune@pragationline.com | www.pragationline.com

Also find us on www.facebook.com/niralibooks

# Preface ...

I am glad to present the book entitled **"Mobile and Wireless Communication"** for **Third Year (Sixth Semester) Diploma in Electronics Engineering** as per SBTE's New Revised syllabus.

I have observed the students facing extreme difficulties in understanding the basic principles and fundamental concepts. To meet this basic requirement of students, sincere efforts have been made to present the subject matter with frequent use of figures.

I am thankful to Publisher Mr. Dineshbhai Furia, Mr. Pradeepbhai Furia, Mr. Jigneshbhai Furia and Staff of Nirali Prakashan specially Mr. P. M. More, Mr. Kiran Velankar, Mr. Ilyas Shaikh, Mrs. Anjali Muley, Mr. Ravindra Walodare for bringing this publication timely to the Market.

I specially thank my students for motivating and encouraging us from time to time.

Any errors or suggestions for the improvement of this book brought to my notice will be thankfully acknowledged and incorporated in the next edition.

**Vijay G. Yangalwar**

# *Syllabus ...*

**Chapter 1 : Introduction to Communication** **[08 Hours]**

A basic cellular system, Performance criteria, Operation of cellular systems, Planning a cellular system, Analog and Digital cellular systems. Examples of Wireless Communication Systems : Paging Systems, Cordless Telephone Systems, Cellular Telephone Systems. Bluetooth and ZigBee.

**Chapter 2 : Elements of Cellular Radio Systems Design** **[08 Hours]**

General description of the problem, Concept of frequency reuse channels, Co-channel interference, Reduction factor, Desired C/I from a normal case in an omni-directional antenna system, Cell splitting, Considerations of the components of cellular systems.

**Chapter 3 : Digital Communication Through Fading Multiple Channels** **[08 Hours]**

Fading channel and its characteristics - Channel modeling, Digital signaling over a frequency, Non-selective slowly fading channel. Concept of diversity branches and signal paths. Combining methods : Selective diversity combining, Switched combining, Maximal ratio combining, Equal gain combining.

**Chapter 4 : Multiple Access Techniques for Wireless Communications** **[08 Hours]**

Introduction, Frequency Division Multiple Access (FDMA), Time Division Multiple Access (TDMA), Spread Spectrum Multiple Access, Space Division Multiple Access, Packet Radio Protocols; Pure ALOHA, Slotted ALOHA.

**Chapter 5 : Wireless Systems and Standards** **[06 Hours]**

AMPS and ETACS, United state digital cellular (IS-54 and IS-136), Global System for Mobile (GSM): Services, Features, System Architecture and Channel Types, Frame Structure for GSM, Speech Processing in GSM, GPRS/EDGE Specifications and Features. 3G Systems : UMTS and CDMA 2000 Standards and Specifications. CDMA Digital Standard (IS 95) : Frequency and Channel Specifications, Forward CDMA Channel, Reverse (CDMA) Channel, Wireless Cable Television.

**Chapter 6 : Future Trends** **[04 Hours]**

4G, 5G and Other Higher G Mobile Techniques, LTE-Advance Systems.

---

✍ ✍ ✍

# Contents ...



✍✍✍

*Chapter* **1**

# INTRODUCTION TO COMMUNICATION

**Syllabus**

A basic cellular system, Performance criteria, Operation of cellular systems, Planning a cellular system, Analog and Digital cellular systems. Examples of Wireless Communication Systems : Paging Systems, Cordless Telephone Systems, Cellular Telephone Systems. Bluetooth and ZigBee.

## 1.1 COMMUNICATION SYSTEMS

### 1.1.1 Introduction

- Man is a social human being. It is his primary need to communicate with each other day-to-day transactions. For short distances, we can talk directly. But for long distances, we cannot talk directly due to attenuation of sound. The word '*Communicate*' is illustrated in the English Dictionary **"as an act of passing news, information, feelings, heat, motion, illness etc."**
- In a broad sense, the purpose of communication is to establish a link between two points. These points may be situated on the earth, or may not be on the earth and the other in space or both in space. A new era establishing a link between two distant points was born with the successful transmission of the first telegraphic message by *Samuel F.B. Morse* in 1836. This is called *electrical communication*. The *electrical communication* means sending, processing and receiving information by electrical means.

### 1.1.2 Definition

- **Communication is the process, whereby the meaningful information is transferred from one point (location) called source in space to the other point (location) called destination (or user).**
- **Communication is a two-way transmission and reception of data streams. Voice, data or multimedia streams are transmitted as signals, which are received by a receiver.**
- The signals from a system can be transmitted through a fibre, wire, or wireless medium. During the transmission process, the transmitter sends the signals according to the defined regulations, recommended standards and protocols.
- **The science of communication involving large distance is called Telecommunication.** The word *'tele'* means *long distance*.
- Now-a-days, satellite band fibre optics have made communication more widespread with an increasing emphasis on *computers* and other *data communications*. Depending upon the type of information to be transferred and received, different electronic communication systems, namely radio telephony, telegraphy, broadcasting, radar, radio telemetry, radio navigation, computer communication, point-to-point and mobile communication systems, such as microwave links etc. have been developed over the years.

### 1.1.3 Types

- Depending on the type of communication used for transmission of electromagnetic signals, the communication systems can be classified into two groups as under :
  1. Wire (or Line) communication.
  2. Wireless (or Radio) communication.

1. **Wire (or Line) Communication :**
   - In the line communication, the mode of transmission is a pair of conducting wires or cables or optical fibres known as *transmission line*. It is also known as *wire communication*.
2. **Wireless Communication :**
   - **Wireless communication is the transfer of information between two or more points that are not connected physically.**
   - Wireless operations permit services such as long range communications that are impossible or impracticable to implement with the use of wires.
   - The wireless communication is also known as *radio communication*, when radio waves are radiated from the transmitter in free space the device called *antenna*.

## 1.2 MOBILE RADIO COMMUNICATION

### 1.2.1 History

- The mobile radio communication has started in the year 1888 after the discovery of *radio* (*electromagnetic*) waves by *Hertz*. Subsequently, *Guglielmo Marconi* demonstrated the *translantic radio telephony* in 1901 and new wireless systems have been developed by people throughout the world.
- The first mobile radio communication system was introduced in the year 1920 for police and emergency services in U.S.A. The first mobile radio communication system was introduced in 1946 for public in U.S.A. This has started the *era of public mobile communication system.*
- The *concept of cellular* was developed by AT and T Bell laboratories of USA in the year 1949. Also in the year 1962, the first test was conducted to explore the commercial applications. With the development of highly reliable, small size, solid state radio frequency (RF) hardware and LSI technology, the *wireless (radio) communication* era was born in the year 1970.
- There was no penetration of mobile phones in the market upto 1980s due to high cost and technical challenges. However, in the last 10 years, the development of wireless (radio) communication systems have been automatically very fast. These developments have helped in making wireless (radio) communication systems smaller, portable and reliable.
- The first digital cellular system was released in the 1990 for the Global System for Mobile (GSM) communication system. The digital switching techniques have facilitated the use of easy to use and affordable communication networks. Now the rate of growth of mobile radio communication has been closely linked with the rate of technological *improvements.*

### 1.2.2 Evolution

- The field of wireless communication is grown rapidly due to the development in the supporting technologies. The growth of mobile communication was slow during the early days and its growth rate has been closely linked with to the rate of technological improvements.
- The first mobile radio communication system was introduced in 1946 for public in U.S.A. after, the discovery of *telegraphy* and started the era of *public mobile communication services*. The cellular concept was first developed by the Bell Laboratories of U.S.A. in 1947 and the first test was conducted to explore commercial applications in 1962.
- The Bell Laboratories proposed to build the first high capacity *cellular telephone system* called Advanced Mobile Public System (AMPS) in 1970. Thus, with the development of highly reliable small size (miniature) and solid. State RF hardware, the era of *wireless communication* has started in 1970.
- The first handheld mobile phone was demonstrated by Dr. Martin Cooper of Motorola company using a handset weighing 2.5 lbs i.e. about 1 kg. The World's first *cellular system* was implemented in 1970 by Nippon Telephone and Telegraph (NTT) company of Japan and the Nordiac Mobile Telephone (NTT 450) was developed in 1981.

- Since the initial commercial introduction of AMPS service in 1988, the mobile radio communication has seen an explosive growth worldwide. The Dyno TAC 8000 was first mobile phone that was made commercially available in 1983. This was the *first generation* (1G) mobile system, which falls under the category of an analog mobile system.
- The *second generation* (2G) mobile system was considered as a digital mobile system. The *third generation* (3G) digital mobile system called the Personal Digital Cellular (PDC) was developed in Japan and is in full commercial operation.
- The worldwide cellular (mobile) phone users increased from 25000 in 1984 to about 25 million in 1993 and from there the growth rate of about 50% per year is being observed. Currently 5G mobile planar are available, which provides very high speed about few Gbps.

### 1.2.3 Introduction

- The mobile communication began in 1897 and evolved remarkably, since when new wireless systems have been developed by people throughout the world. The development of wireless systems has been dramatically fast in the past 10 years or also.
- The improvements in the RF and digital fabrication technologies, the new large scale circuit integration (LSI) and other miniaturization techniques have helped the development of wireless to a great extent. These developments have helped in making smaller portable and more reliable radio equipments. The digital switching techniques have fascilated the use of a way to use and affordable communication networks.

### 1.2.4 Concept

- **The satellite communication enabling as to provide universal personal communications with terminal mobility, personal mobility and service mobility is called mobile communication.**
- The concepts enabling as to provide universal personal communications are as given below :
  1. Terminal mobility.
  2. Personal mobility.
  3. Service mobility.
- These concepts are being catalyzed at the national, regional and international levels to specify and standardize a range of mobile and personal communication systems and services. The basis for *terminal mobility* and *personal mobility* is the use of wireless access and personal (rather than terminal) numbers respectively. *Service mobility* is based on the emerging Intelligent Network (IN) concepts, which facilitate real time management service profiles. It is shown in Fig. 1.1.

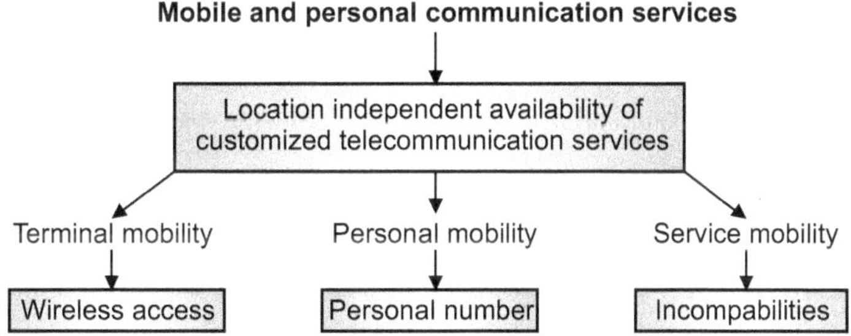


**Fig. 1.1 : Concepts of mobile and personal communication services**

### 1.2.5 Frequencies Used

- For cellular communications, the FCC has proposed 40 MHz of the frequency spectrum ranging from 825 MHz to 845 MHz and 870 MHz to 890 MHz. Full duplex operation is possible by separating transmit and receive signals into separate frequency bands.

- Cellular phone units transmit in the lower bands of frequencies 825 MHz to 845 MHz and receive in the higher band of 870 MHz to 890 MHz. The opposite frequency bands are used by the base units of the cell sites. Within these two bands, 666 separate channels (333 channels per band) have been assigned for voice and control. Each channel occupies a bandwidth of 30 KHz.

### 1.2.6 Advantages

1. It can offer a large variety of data services and supplementary services tailored to system type.
2. It has a high level of security such as authentication, encryption etc.
3. It has a high traffic density and provides more than 500 channels.
4. It has a high coverage range and area.
5. It has a good speech quality with minimized noise.
6. It can be used in any location such as home, office, in public and in transit.
7. It can be used by using devices such as cellular/PCs phone, office wired phone, Personal Digital Assistant (PDA), home phone, fax, multimedia terminal etc.
8. It provides seamless global roaming and improved quality of services to its users.
9. It is compatible with ISDN leading to wider range of services.

### 1.2.7 Applications

1. Paging and short message services.
2. E-mail and Internet/Intranet access.
3. Database access and file transfer.
4. Fleet management and dispatch (tracks, taxis, packages delivery systems etc.).
5. Inventory control (stores, warehouses etc.).
6. Field services.
7. Intelligent highways (e.g. automatic toll collection).
8. Data inputs e.g. hospital patient charts, auto rental returns etc.
9. Credit card authorization (from remote or mobile locations).

## 1.3 WIRELESS COMMUNICATION SYSTEMS

### 1.3.1 Introduction

- In 1897, *Guglieno Marconi* was the first to demonstrate that it was possible to establish a continuous communication system with the ship.
- The present days, wireless communication system has become an essential part of various types of communication devices that permits user.

### 1.3.2 Definition

- **Wireless communication is a type of data communication, that is, performed and delivered wirelessly.**
- This is a *broadband* term that incorporates all procedures and forms of connecting and communicating between two or more devices using a wireless signal through wireless communication technologies and devices. The communication or receiving intermediate device captures these signals, creating wireless communication bridge between the sender and receiver device.
- **Any transfer of information between points that do not have a physical connection, like wire, cable, etc. is called wireless communication.**

### 1.3.3 Practical Examples

1. Garage door openers.
2. Hand-held Walkies-Talkies.
3. Remote controllers.
4. Paging systems (Pager).
5. Cordless telephones.
6. Cellular telephones.

- However, the cost, complexity, performance and types of services offered by each of these mobile systems are vastly different.

## 1.4 RADIO PAGING SYSTEMS

### 1.4.1 Definition

- **The pager is a specialized miniature radio receiver carried by the user for retrieval of information from the paging terminal.**
- Depending on the *pager* and *system*, the user can receive either an alert, a voice transmission, a numeric display page, or an alpha numeric display page.
- **A pager is a wireless telecommunications device that receives and display alpha numeric or voice messages.**
- One-way pager can only receive messages, while response pagers and two-way pagers can also acknowledge, replay to and originate messages using an internal transmitter.

### 1.4.2 Block Diagram

- The block diagram of a paging system is as shown in Fig. 1.2

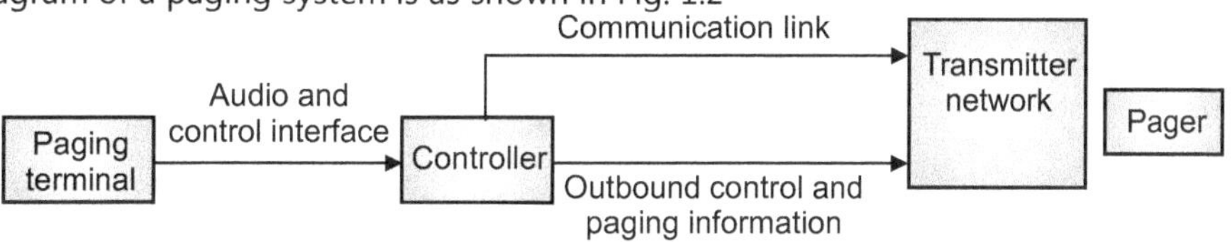


**Fig. 1.2 : Functional block diagram of a radio pager system**

- A paging system is composed of the following six basic elements or components :
  1. Paging terminal,
  2. Controller,
  3. Outbound control and paging information channel,
  4. Transmitter network,
  5. Communications link
  6. End user pagers.

### 1.4.3 Description

The function of each block of a radio paging system is described as under :

1. **Paging Terminal :**
   - The paging terminal maintains subscriber pager type and service information. The terminal accepts the paging requests via direct dial-up phone interface or via data entry equipment. The terminal prepares the paging information for transmission and communicates to the controller that it needs access to the infrastructure.
   - Terminals can also provide statistics on paging traffic and billing information. The paging terminal is connected to the controller for key requests and control handshakes as well as passing pager address and information for transmission. The terminal is, therefore, co-located to allow for control signal and audio interface.
   - The paging terminal provides user access typically by a Video Display Terminal (VDT) for database programming and review purposes. Terminals can usually support a variety of peripherals, such as disk drives, printers, and modem for expanded user capabilities.

**2. Controllers :**

- The controller is the control focal point of the system responsible for recognizing paging terminal requests and keying the transmitters in the proper paging mode, analog or binary. The controller depending on its sophistication is responsible for monitoring transmitter operational performance as well as reporting system level alarms.
- The controller can also perform system level maintenance to compensate for changes in the out bound control and paging information channel. Depending on the type, the controller may or may not store transmitter configuration and operational data, as well as logging station and system alarm information.
- The controller provides user access typically by a Vide Display Terminal (VDT) for database programming and review. More sophisticated controller usually supports a variety of peripherals such as disk drives, printers, and modems for expanded user capabilities.

**3. Outbound Control and Paging Information Channel :**

- The outbound control and paging information channel is the medium by which the controller information is passed on to the transmitters. The most conventional information distribution methods are RF links, dedicated wire lines to each transmitter from the controller site, or satellite up-link and down links. With an RF link system the controller audio is sent to a link transmitter via a wire line connection.
- A link repeater control then be used to listen to the link transmitter to further extend transmission of the information, hence the name repeater. With a wire line system the controller audio is sourced to a data splitter, which then feeds the signal to each transmitter via a dedicated wire line. With a satellite distribution system, the controller information is sent to the satellite up-link location. Each station is equipped with a satellite dish and receiver the recover the controller information.

**4. Transmitter Network :**

- The paging base stations decide control information from the controller for keying in the appropriate mode, analog or binary. The transmitter network in conjunction with the antenna system converts the information from the paging terminal into modulation and RF energy for transmission to the paging receivers.
- The transmitters continually check their operating performance and generate alarms in the event of a degradation in performance. The controller and system configurations determine how and when the alarms are reported.

**5. Communications Link :**

- The communications link is primarily feedback path from the transmitter network to the controller. This communications link is used to transfer the base station status and alarm information from the transmitter network to the controllers for evaluation and reporting. Several communications links may exist from the transmitter network to the controller.
- The communications link can be a one way path from base station to controller or a bi-directional depending on the type of link used. Some common communications links include dedicated wire lines, dial-up plane links, and monitor receivers with dedicated return phone lines.

**6. Pager :**

- The pager is a specialized miniature radio receiver carried by the user for retrieval of information from the paging terminal. Depending on the pager and system, the user can receive either an alert, a voice transmission, a numeric display page, or an alpha-numeric display page.
- The pager is only able to receive pages within the allowable coverage area provided by the transmitter network.

### 1.4.4 Messages to Paging Systems

- One of the following four types of information messages can be delivered in a paging system.
  (1) Alert tone message
  (2) Voice message
  (3) Digital string message
  (4) Text string message

### 1.4.5 Applications

(1) Paging systems are typically used to notify a subscriber of the need to call a particular telephone number or travel to a known location to receive further instructions.

(2) The modern paging systems are used to send News headlines, Stack quotations, and Faxes.

## 1.5 CORDLESS TELEPHONE SYSTEMS

### 1.5.1 Brief History

- The cordless phones first appeared around 1980. The earliest cordless phones operated at frequency of 27 MHz. They had the following problems :
  1. Limited range
  2. Poor sound quality
  3. Poor security.
- In 1986, the Federal Communications Commission (FCC) granted the frequency range of 47-49 MHz. For cordless phone, which improved their interference problem and reduced the power needed to run them. However, the phones still had a limited range and poor sound quality. Because the 43 - 50 MHz cordless phone frequency was becoming increasingly crowded, the FCS granted the frequency range of 900 MHz in 1990.
- This higher frequency allowed cordless phones to be clearer, broadcast a longer distance and chooses from more channels.
- In 1994, digital cordless phones in the 950 MHz frequency range were introduced. Digital signals allowed the phones to be more secured and decrease of eave dropping, it was pretty easy to eave drop an analog cordless phone conversations.
- In 1995, digital spread spectrum (DSS) was introduced for cordless phones. This technology enable of the digital information to spread in pieces over several frequencies between the receiver and the base, thereby making it almost impossible to eaves drop on the cordless conversations. In 1998, the FCC opened up the 2.4 GHz range for cordless phone use. Thus, frequency has increased the distance over which a cordless phone can operate and brought it out of the frequency range of most radio scanners, thereby further increasing security.

### 1.5.2 Concept

- Cordless Telephone as portable telephone is a telephone with a wireless handset that communicate via *radio waves* with a base station connected to a fixed telephone line, usually within a limited range of its base station.
- A cordless telephone a basically a combination of telephone and radio transmitter/receiver.

### 1.5.3 Features

1. Speaker phone
2. Caller ID
3. Mail box
4. LCD screen
5. Headphone jack
6. Two line support
7. Auto talk.

### 1.5.4 Block Diagram

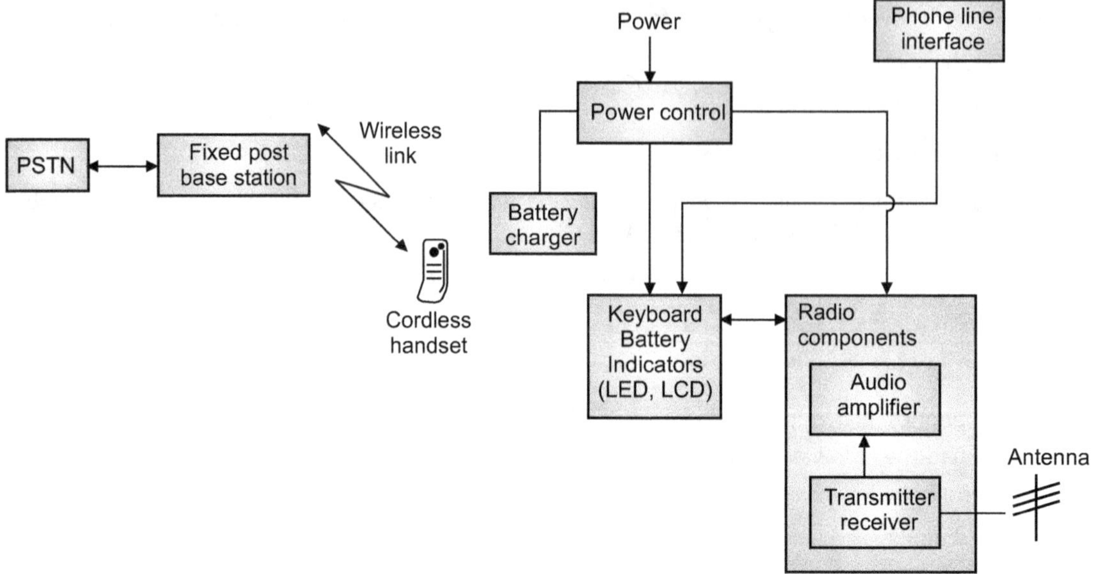


**Fig. 1.3 : Block diagram of a cordless system**

- The block diagram of a cordless phone system is as shown in Fig. 1.3. It consists of a wireless network, paging system and cellular system.

### 1.5.5 Basic Principle

- The dedicated base station is then connected to a dedicated telephone line with a specific telephone number on the Public Switched Telephone Network (PSTN). In first generation cordless telephone systems (CT 1) manufactured in the 1980s, the portable unit communicates only to the dedicated base unit and only over distance of a few tens of meters.
- Early cordless telephones operate solely as extension telephones to a transreceiver connected to a subscriber line on the PSTN and are primarily for in-home use. CT 2 is a second-generation cordless telecommunication system. The digital cordless standards are designed for wireless PABX and key systems, cordless residential telephones and public cordless access (telepoint) services.
- The CT 2 standard originated in the U.K. for providing public telepoint services and was adopted in the year 1992 by ETSI as an interium European standard for cordless communications. The CT 2 plus standard, adopted by Canada, is a fully compatible with CT 2 and provides as standard features more efficient location tracking mechanisms, and faster incoming call delivery and handoff.

### 1.5.6 Working

- A cordless phone has two major parts : (1) Base and (2) Handset. The function of each block of a cordless telephone system is described as under :

1. **Base :**
   - The base is attached to the phone jack through a standard of phone wire connection, it looks just like a normal phone. The base receives the incoming call as an electrical signal through the phone line, converts it to an FM radio signal then broadcast that signal.
2. **Handset :**
   - The handset receives the radio signal from the base. Converts it to an electrical signal and sends that signal to the speaker, where it is converted into the sound that you hear. When you talk, the handset broadcasts your voice through a second FM radio signal back to the base. The base receives your voice signal, converts it to an electrical signal and sends that signal through the phone line to the other party.
   - The base and handset operate on a frequency pair that allows you to talk and listen at the same time, called *duplex* frequency.

### 1.5.7 Operating Frequencies

- Cordless phones operate in frequencies mandated byy the FCC. There are four major frequency band options as under :

  (i) 43 - 50 MHz (ii) 900 MHz

  (iii) 2.4 GHz and (iv) 5.8 GHz

### 1.5.8 Application

- The main application environments for CT 2 include residential cordless setups, public telepoint systems and wireless PBX and key systems.

## 1.6 CELLULAR TELEPHONE SYSTEMS

### 1.6.1 Introduction

- Cellular telephone systems, also referred to as Personal Communication Systems (PCs), are extremely popular and lucrative world wide. These systems are designed to provide two-way voice communication at vehicle speeds with regional or national coverage.
- Cellular systems were initially designed for mobile terminals inside vehicles with antennas mounted on the vehicle roof. The basic feature of the cellular system is frequency reuse, which exploits path loss to reuse the same frequency spectrum at spatially separated locations.
- To reduce the high cost of base stations, the cellular systems use a relatively small number of cells to cover an entire city or region. Cellular telephone systems have moved from analog to digital telephony because digital technology has many advantages. These systems provide voice mail, paging, and e-mail services in addition to voice.
- Digital cellular systems can use any of the multiple access techniques. The first cellular telephone for commercial use was approved by the Federal Communications Commission (FCC) in 1983.

### 1.6.2 Concept

- Cellular telephony is a system level concept which replaces a single high power transmitter with a large number of low-power transmitters for communication between.
- A cellular telephone system provides a wireless communication over large geographical area.
- The cellular system uses the concept of frequency reuse, which exploits the path loss to reuse the same frequency.

### 1.6.3 Definition

- **Cellular telephone, sometimes called mobile telephone is a type of short-wave analog or digital telecommunication, in which a subscriber has a wireless connection from a mobile phone to a relatively nearby transmitter.**
- The transmitter span of coverage is called a *cell*. As the cellular telephone user moves from one cell or area of coverage to another, the telephone is effectively passed on to the local cell transmitter.
- A cellular telephone is not to be confused with a cordless telephone. A cellular telephone system provides a wireless connection to the PSTN for any user location within the radio range of the system.

### 1.6.4 Overview Diagram

- The overview diagram of a cellular telephone system is as shown in Fig. 1.4.

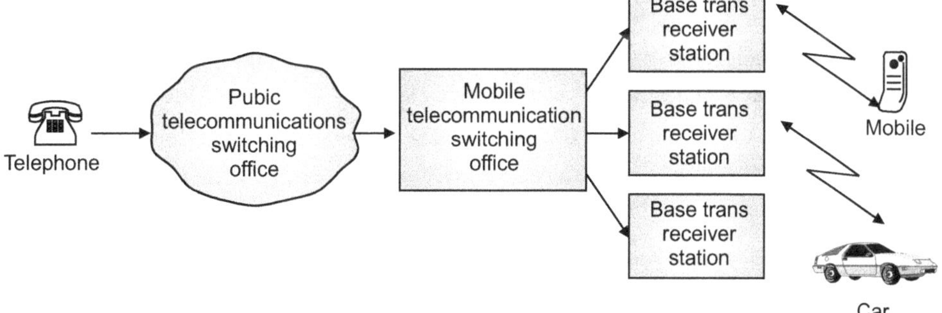


**Fig. 1.4 : Overview of a cellular telephone system**

### 1.6.5 Block Diagram

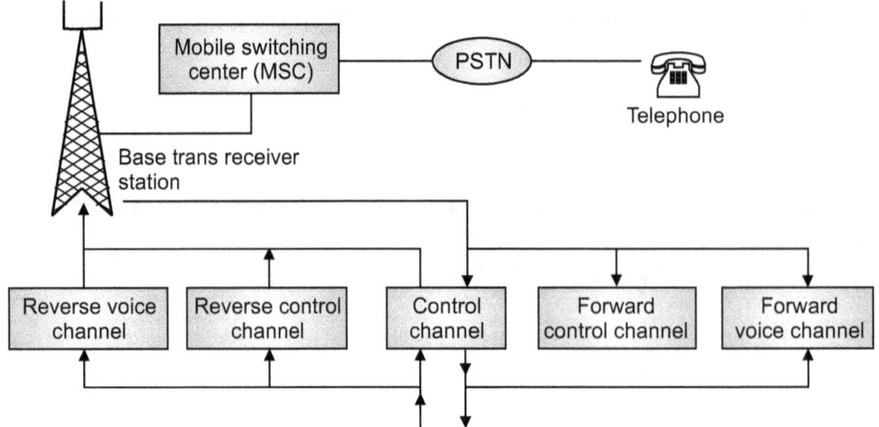


**Fig. 1.5 : Block diagram of cellular telephone system**

- The block diagram of a cellular telephone system is shown in Fig. 1.5.
- It consists of the following components :

  (1) Mobile stations, (2) Base transceiver station, (3) Mobile Switching Centre (MSC).

## 1.7 WIRELESS LOCAL LOOP (WLL)

### 1.7.1 Introduction

- **Wireless Local Loop (WLL) is a generic term for an access system that uses wireless link rather than conventional copper wires to connect subscribers to the local telephone company's switch.** It is also known as fixed wireless access (FWA) or simply fixed radio.
- This type of system uses analog or digital radio technology to provide telephone, facsimile, and data services to business and residential subscribers. WLL systems provide rapid deployment of basic phone service in areas, where the terrain or telecommunications development makes installation of traditional wirelines service too expensive.
- WLL systems can be easily integrated into a wireline Public Switched Telephone Network (PSTN) for more quickly than the traditional wireline installations.

### 1.7.2 Definition

- Wireless Local Loop (WLL) is the use of a wireless communications link as the connection for delivering Plain Old Telephone Service (POTS) or Internet access marked under the term 'broadband' to telecommunications customers.
- **Wireless Local Loop (WLL) is a protocol used to connect a user to Public Switched Telecommunications Network (PSTN) or broadband internet using RF signals.**

### 1.7.3 Need

- The rapid growth of the internet has created great demand for *broadband* internet and computer access from business and residential throughout the world. This needs a single broadband internet connection that could provide all the telecommunications services, e.g. telephone, television, radio, facsimile, internet etc., for home or business customers. The WLL network is extremely well suited for the broadband applications. Therefore, WLL network is very important for broadband internet applications. Therefore, there is a need of WLL network for broadband internet applications.

### 1.7.4 Features

1. Better quality of service.
2. Compatible with other cellular technologies.
3. Scalability.

### 1.7.5 WLL Architecture

- The architecture of Wireless Local Loop (WLL) is as shown in Fig. 1.6.

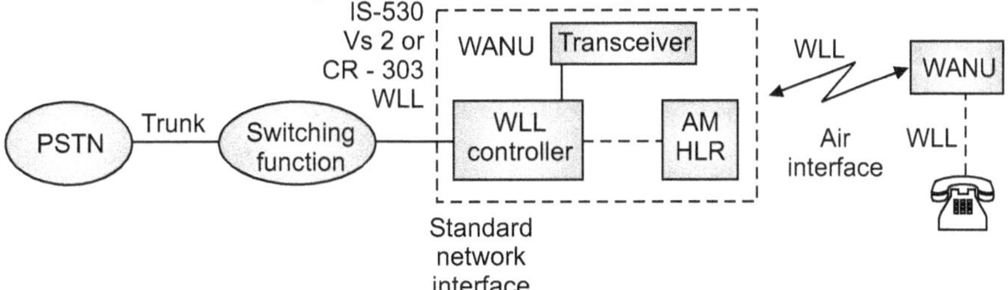


**Fig. 1.6 : WLL reference model architecture**

- A simplified version of the TR-45.1 architectural reference model for WLL is as shown in Fig. 1.6. In this figure, the wireless access network unit (WANU) consists of the base station transceivers (BTSs) or radio port (RP), the radio controller (RPCD), an access manager (AM) and home location register (HLR). The interface between the WANU and the switch is called WLL that can be IS-634, IS-653, GR-303, VS-1, ISDN.

### 1.7.6 Advantages

1. The cost of wireless equipment is less.
2. Easy installation.
3. They have good scale of installation.

### 1.7.7 Applications

1. Local phone service via wireless connection.
2. Cheaper to install than wired lines.
3. Very prominent in non-industrialized nations.

## 1.8 PLANNING A CELLULAR SYSTEM

### 1.8.1 Introduction

- Cellular technology has rapidly evolved as the prime form of wireless communication over the last decade. The wide-spread deployments of cellular network derive motivation from the need to provide mobile telephony service, so as to enhance the capacity in terms of the number of users.
- However, radio spectrum is a scarce resource, which inhibits the indefinite growth of user based capacity. As a consequence, the radio frequencies have to be efficiently reused numerous times for several thousand transreceivers by exploiting the inherent property associated with propagation of radio waves that the signal strength decreases with increase in distance.
- The reuse takes place after a safe minimum distance at which cells having same frequencies (co-channel) do not interfere. Additionally, no two neighbour cells/sites use adjacent frequencies to avoid in-band interferences. In order to satisfy the cellular frequency reuse principle, frequency planning is required. In this regard, the concept of manual frequency planning and inability to fulfill the conditional interdependencies for a large number of transmitter sites.

### 1.8.2 Definition

- In the context of mobile radio communication systems, planning means the planning of radio frequency. **RF planning is the process of assigning radio frequencies, transmitter locations and parameters of a wireless communications system to provide sufficient coverage and capacity for the services required.**
- The RF plan of a communication system has two objectives : (1) *Coverage* and (2) *Capacity*. Coverage relates to the geographical footprint within the system that has sufficient RF signal strength to provide for a call/data session. Capacity relates to the capability of the system to sustain a given number of participants. Capacity and coverages are interrelated. To improve coverage, capacity has to be sacrified, while to improve capacity, coverage will have to be sacrified.
- **Frequency planning in ever growing cellular networks is an extremely arduous task.**

- The automated frequency planning system is based on the availability of a reliable prediction tool. Frequency planning offers a solution to avoid *interference* caused by the intermodulation planning and inability to fulfill the conditional interdependencies for a large number of transmitter sites.

## 1.8.3 Construction

- The RF planning process consists of following four major stages :
  1. Initial radio link budgeting.
  2. RF propagation modeling.
  3. Fining, tuning and optimization.
  4. Continuous optimization.

**1. Initial radio link budgeting :**

- The first level of the RF planning process is a budgetary level. It uses the RF link budget along with a statistical propagation model to approximate the coverage area of the planned sites and to eventually determine how many sites are required for the particular RF communication system.
- The statistical propagation of the model does not include *terrain* effects and has a slope and intercept value for each type of environment, e.g. Rural, Urban, Suburban, etc. Two essential inputs at this level are simple radio transreceiver characteristics and flat map of the area.
- This fairly simpliest approach allows for a quick analysis of the number of sites that may be required to cover a certain area.

**2. RF propagation modeling :**

- The second level of RF planning process relies on a more dedicated propagation model. Automatic planning tools are often employed in this phase to perform detailed predictions. The propagation model takes into account the characteristics of the selected antennas, the terrain, and the land use and land clutter surrounding each site.
- This requires a precise and accurate characterization of every transreceiver and detailed, three dimensional model of the *terrain*. Since these factors are considered, this propagation model provides a better estimate of the coverage of the sites that the initial statistical propagation model.
- Thus, its use in conjugation with the RF link budget, produces a more accurate determination of the number of sites required. Following is a typical list of outputs produced at this stage.
  1. Number of sites.
  2. Site locations and height.
  3. Antenna directions and down tilts.
  4. Neighbouring cell lists for each site.
  5. Mobility.
  6. Frequency plan.
  7. Detailed coverage predictions.

**3. Fining, tuning and optimization :**

- The third phase of the RF planning process incorporates further detail into the RF plan. This stage included items, such as collecting drive data to be used to tune or calibrate the propagation prediction model, predicating the available data throughout each site, fine tuning of parameter settings, e.g. antenna orientation, down tilting and frequency plan.
- This process is required in the development of the system or in determining service contract-based coverage. Following is a typical list of outputs produced at this stage.
  1. Number of sites.
  2. Site locations and height.
  3. Antenna directions and downtilts.
  4. Neighbour cell lists for each site.
  5. Mobility.
  6. Frequency plan.
  7. Detailed coverage prediction.

**4. Continuous optimization :**

- The final phase of the RF planning process involves continuous optimization of the RF plan to accommodate for changes in the environment or additional service requirements, e.g. additional coverage or capacity.

- This phase starts from initial network deployment and involves collecting, measuring data on a regular basis that could be via drive testing or centralized collection. The data is then used to plan new sites or to optimize the parameter settings, e.g. antenna orientation, downtilting, frequency plan of existing sites.

## 1.9 ANALOG AND DIGITAL CELLULAR SYSTEMS

### 1.9.1 Analog Cellular System

- Nippon Telegraph and Telephonic Corporation (MTT) developed an 800 MHz land mobile telephone system and put it into service in the Tokyo area in 1979. The general system operation is similar to the AMPs system. It accesses approximately 40,000 subscribers in 500 cities. It covers 75% of all Japanese cities, 25% inhabitable areas, and 60% of the population. In Japan, 9 automobile switching centers (ASCs), 51 mobile control stations (MCSs), 163 mobile base stations (MBSs), and 39,000 mobile subscriber stations (MSSs) were in a position as of February 1985.
- The Japanese mobile telephone service network configuration is used. In the metropolitan Tokyo area, about 30,000 subscribers are being served. The 1985 system operated over a spectrum of 30 MHz. The total number of channels was 600 and the channel bandwidth was 25 kHz.
- This system comprised an ASC, MCS, a MBS. As there is no competitive situation set-up by the government. However, the Japanese Ministry Post and Telecommunication (MPT) is considering providing a dual competitive situation similar to that in United States.
- Nordic system (NMT) was built most by Scandinavian countries (Denmark, Norway, Sweden, and Finland) in cooperation with Saudi Arabia and Spain and is called the NMT network. It is currently a 450 MHz system. But on 800 MHz system will be implemented soon, since the frequency transport concept as the AURORA 800 system is used to convert the 450 MHz system to the 800 MHz system.
- The total bandwidth is 10 MHz, which has 200 channels with a bandwidth of 25 MHz per channel. This system does not have **landoff** and **roaming** capabilities. It also uses repeaters to increase the coverage in traffic area. The total number of subscribers is around 100,000.
- In 1945, the Zero Generation (0G) of mobile telephones was introduced. They were not **cellular** and did not have the feature of **handover**, reuse of RF and **hand-off**. The Nordic Mobile Telephone (NMT) system went on-line in Denmark, Finland, Norway and Sweden in 1981. Personal **handy phone** system and **modems** used in Japan around 1997 - 2003.
- In 1983, Motorola DynoTAC was the first approved mobile phone by FCC in the United States. In 1984, Bell laboratories developed modern commercial cellular technology by which employed multiple, centrally controlled base stations (cell site), each providing service to a small area (a cell). Cellular systems such as AMPS, TACS, NMT are analog.

### 1.9.2 Limitations of Analog Systems

1. Limited service capability.
2. Poor service performance.
3. Inefficient frequency spectrum utilization.

### 1.9.3 Performance Criteria

1. Voice quality
2. Service quality
3. Special features
   (i) Call forwarding
   (ii) Call waiting
   (iii) Voice Sourced Box (VSR)
   (iv) Automatic roaming
   (v) Navigation services.

# 1.10 COMPARISON OF ANALOG AND DIGITAL CELLULAR SYSTEMS

**Table 1.1**

| Parameter | Analog Cellular System | Digital Cellular System |
|---|---|---|
| 1. Traffic channel | Voice use FM. | Voice encoded in digital format. |
| 2. Processing | More difficult. | Easier. |
| 3. Encryption | No security. | Yes, security. |
| 4. Noise | More noisy. | Less noisy. |
| 5. Error detection and correction | No such facility. | Yes, such facility available. |
| 6. Channel access | One channel to only one user. | One channel shared by number of users. |
| 7. Compatibility | No compatibility. | Compatibility with computers. |

# 1.11 MOBILE RADIO STANDARDS

## 1.11.1 Introduction

- Digital Mobile Radio (DMR) is an *international open digital mobile radio standard* defined in the European Telecommunications Standards Institute (ETSI) standards TS 102 361 parts 1 - 4 and used in commercial products around the word.
- DMR, Digital Mobile Radio is an *international digital radio standards* specified for business mobile developed by ETSI, European Telecommunications. In short, DMR is a digital two-way radio standard offered by ETSI for global use.

## 1.11.2 Types

- Some of the important world-wide mobile radio standards are as under :
  1. AMPS
  2. NAMPS
  3. IS-95
  4. GSM
  5. DMTS
  6. CDMA-2000
- All these mobile radio standards are developed in North America.

## 1.11.3 AMPS Standards

- The Advanced Mobile Phone System (AMPS) based on cellular concept was developed by Bell Laboratories in Chicago (U.S.A.) in the year 1977-78. The world's first cellular system was implemented by Nippon Telephone and Telegraph Company (NTT), Japan in the year 1978.
- The cellular mobile service licences for an initial 40 MHz spectrum in the 800 MHz frequency band were issued and subsequently another 10 MHz was added. Thus, the current spectrum allocation for cellular mobile radio in the U.S.A. is 50 MHz within the 824 to 849 MHz (uplink) and 869 to 894 MHz (downlink) frequency bands. In a given licencing area, the spectrum is shared by two operators, the Wireline Common Carrier (WCC) and the Radio Common Carrier (RCC).
- The WCC is an arms-length subsidiary of a Local Exchange Carrier (LEC), which provides local wired telephone service in the licencing area. Under the AMPS standards, which specifies a carrier spacing of 30 kHz, the 50 MHz spectrum leads to a total of 832 full duplex channels with 416 channels each for the A-band (RCCs) and the B - band (WCCs) operators in each licencing area.
- Out of these 416 channels, 21 channels are used as control or set up channels, with the remaining 395 channels for user traffic. The later are then grouped and each group is assigned to cells within a cluster.
- Before digital cellular systems based on Time Division Multiple Access (TDMA) and Code Division Multiple Access (CDMA) technology came on the scene, the AMPS system dominated cellular implementations around the world.

- The AMPS services were commercially introduced in U.S.A. in the year 1983. Since then the mobile communication has seen an explosive growth world wide. The European Total Access Cellular System (ETACS) was developed in the year 1985. The American system AMPS was almost identical to European System ETACS. The earlier European system was not compatible to another because of different frequencies used and different communication protocols being used by them.

## 1.11.4 N-AMPS Standards

- The North American Mobile Phone System (N-AMPS) was frequency modulation with 12 kHz deviation for speech (8 kHz for signaling). By adopting the wider 12 kHz deviation for a 30 kHz channel spacing, the AMPS system provides an external dynamic range for speech and therefore increased protection against cochannel interference.
- This measure combined with the use of speech companders, yields a high quality voice channel with the capability to maintain good cochannel performance in a high capacity configuration. The signaling between the *mobile station* and the *base station* is at 10 kbps, with Manchester encoding, the bit rate is extended to 20 kbps.
- The data on the signaling channel are modulated on the radio carrier using direct frequency shift keying with a peak frequency deviation of 3 kHz. *Bose-Choudhary (BCH) encoding* is used to protect the signaling data from multipath fading. Besides the data transmission on the signaling channel for call set up, data also are transferred on the speech or voice channel, a *blank-and-burst* technique is used, whereby the voice signal is balanced for about 50 ms and a data burst of 10 kbps is inserted in the voice channel.
- This signaling is used for such features as alerting the mobile about an impending channel transfer for a hand-off. To meet the co-channel interference objectives, the typical frequency reuse plan employed in AMPS systems is either 12 group frequency cluster with omni-directional antennas or a 7 group cluster with three sectors per cell. Further, while the call is in progress, the base station transmits a low level *Supervisory Audio Tone* (SAT) in the region of 6 kHz.
- Three different SAT frequencies are used by the network and are allocated to the base station such that base stations most likely to cause the interferences have a different SAT from the serving base station. The mobile continuously monitors the received SAT and also transponds the signal back to the serving base station. If the mobile (or the base station) detects a difference between the received SAT and the one expected, the audio path is muted to prevent the interfering signal from being overheard. If the condition persists, the call may be aborted.
- A narrow band version of AMPS is called N-AMPS. It utilizes 10 kHz channel (instead of 30 kHz) has also been standardized by Telecommunication Industry Association (TIA) of USA. The motivation for the N-AMPS was to provide a three-fold increase in capacity. A number of systems based on the NAMPS standard have been implemented and are in commercial operation.

## 1.11.5 IS-95 Standards

- The CDMA offers many advantages over FDMA and TDMA. So, a digital cellular system based on CDMA was standardized in U.S.A. as Interim Standard-95 (IS-95) by Telecommunication Industry Association (TIA). The IS-95 system uses CDMA standards. The CDMA assigns a unique code to each user. So every, user is allowed to access the channel fully all the time. It is designed to compatible with the existing AMPS system frequency band.
- In IS-95, each user within a cell is allowed to use the same radio channel and the user in the adjacent cells also use the same radio channel because this system uses the direct sequence spread spectrum CDMA. The user data rate changes in real time and it depends on the voice activity and requirements of the network. In IS-95, different modulation and spreading techniques are used for the forward and reverse links. The CDMA systems based on IS-95 standard are now in commercial operation in North and South America as well as in Japan, Korea and China.

### 1.11.6 GSM Standards

- The GSM stands for Global System for Mobile. It is one of the most popular mobile communication standard and the GSM communication uses the cellular networks and TDMA. It is a second generation (2G) cellular mobile radio system standard and it was the world's first digital cellular system using digital modulation.
- The GSM standard was developed by European group called Conference of European Post and Telecommunication (CEPT) Administration in the year 1990. The first GSM system was implemented in Germany in the year 1992 and it was named as $D_2$. The most important service supported by GSM is telephony. Other services derived from telephony included in the GSM specification are emergency calling and voice messaging.
- In GSM, the uplink (mobile to base) frequency band is 890 to 915 MHz and the corresponding downlink (base to mobile) is 935 to 960 MHz; resulting in a 45 MHz spacing duplex operation. The available 25 MHz spectrum is partitioned into 124 carriers (carrying spacing of 200 MHz) and each carrier is divided into 8 time slots (radio channels).

### 1.11.7 UMTS Standards

- The Universal Mobile Telecommunications System (UMTS) is a third generation (3G) mobile cellular system for network based on the GSM standard. It is developed and maintained by this 3G Partnership Project. It is a component of the International Telecommunications Union IMT-2000 standard set for network based on the computing CDMA-1 technology.
- It is a 3G mobile cellular system that is capable to provide a variety of mobile services to a wide range of GSM standards. It uses wideband code division multiple access (W-CDMA) radio access technology to offer greater network operators.
- It specifies a complete network system, which includes the radio access network (UMTS) Terrestrial Radio Access Network or UTRAN), the core network (Mobile Application Part as MHP) and the authentication of users via SIM (Subscriber Identity Mobile) cards. The technology described in UMTS is sometimes also referred to as Freedom of Mobile Multimedia Access (FDMA) or 3 GSM.

### 1.11.8 CDMA-2000 Standards

- CDMA-2000 is a family of 3G mobile technology standards for sending voice, data, and signaling data between mobile phones and cell sites. It is a 3G mobile wireless technology. CDMA-2000 specification was developed by the third generation (3G). It is basic 3G standard, but in what is termed as CDMA-2000 1X EV. There are further developments. There are basically only two types. It represents a family of IMT-2000 (3G) standards providing the high quality voice and broadband into services over wireless networks.
- It is an upgradation of existing 2G and 2.5 CDMA technology and it can support much higher data rates than that of 2G and 2.5 G system. The basic structure of 3G CDMA-2000 is same as that of 2G CDMA system with a channel band width of 1.25 MHz per radio channel.

## 1.12 BLUETOOTH TECHNOLOGY

### 1.12.1 Introduction

- Bluetooth is a well-known short-range technology for Wireless Personal Area Networks (WPAN).
- Starting from a headset cable replacement it has been extended to support flexible ad-hoc networks. To extend from low bit rate data to streaming multimedia, Quality of Service (QoS) is required.
- Bluetooth specification defines strong interoperability demands between all bluetooth devices. The interoperability requirements demand a lot from application developers.
- For making the developer's work easier it has been produced different bluetooth development platforms. These development platforms have different purposes and capabilities.
- Development platforms make the development work easier, but many tools are not compatible with the bluetooth specifications.

- Bluetooth wireless specification got its name from the 10th century Danish king, who used diplomacy to negotiate a truth between two finding fractions.
- Bluetooth is an always on, short range radio hook up that resides on a microchip. It was initially developed by Swedish mobile phone maker Ericsson in 1994 as a way to let laptop computers make calls on a mobile phone.
- Since then several thousand companies have signed on to make Bluetooth the low-power short-range wireless standard for a wide range of devices.
- The bluetooth is a wireless technology used for transferring the data from one device to the other device. The distance between two devices is very short from the fixed mobile device and building area network.
- The bluetooth technology was developed by the Bluetooth special interest group and its physical range is from 10 m to 10 cm i.e. about 20 feet to 300 feet.
- The bluetooth device can connect upto seven devices and it is used in the industry like smartphones, personal computers, and gaming consoles etc. The IEEE standardized Bluetooth as IEEE 802.15.1 but the standards are maintained for short periods.
- The bluetooth technology does not require any cables, cords and adopters to communicate with other devices.

## 1.12.2 Concept

- The concept behind Bluetooth is to provide a universal short-range wireless capability.
- Using the 2.4 GHz band, available globality for unlicensed low-power user, two bluetooth devices within 10 meter of each other can share upto 720 kbps capacity.
- Bluetooth is intended to support an open-ended list of applications, including data such as Schedules and telephone numbers, audio, graphic and even video. For example, audio devices can include headsets, cordless and standard phones, home stereos and digital MP3 players.

## 1.12.3 Definitions

- The bluetooth technology is a wireless technology standard, which is used to transfer the data between two different electronic devices over short distance.
- The bluetooth technology is a high-speed, low-powered wireless technology link that is designed to connect phones or other portable equipments together.
- The bluetooth technology is a short-range wireless communications technology to replace the cables connecting electronic devices, allowing a person to have a phone conversation via a headset, use a wireless mouse and synchronized information from a mobile phone to a personal computer.

## 1.12.4 Features

1. It has capability of sharing all its features with other bluetooth devices in the surrounding area.
2. It can share information with a bluetooth enabled computer or printer.
3. It can share all the features to internet.
4. It can communicate at ranges upto 10 meters.
5. It does not need to be in direct sight of each other.
6. Less complication
7. Less power consumption
8. Available at cheaper rates
9. Robustness
10. Hands free headset
11. Uniform structure
12. Global acceptance
13. Interactive conference

## 1.12.5 Classification

1. Bluetooth headsets
2. Stereo headset
3. In-car bluetooth headset
4. Bluetooth equipped printer
5. Bluetooth enable webcom
6. Bluetooth keyboard
7. Bluetooth GPS device

### 1.12.6 Basic Principle

- Bluetooth is a connective convenience. It is a high-speed, low-power microwave wireless link technology, designed to connect phones, laptops, PADs and other portable equipments together with little or no work by the user.
- Unlike infra-red, Bluetooth does not require **line-of-sight** positioning of connected units.
- The technology uses modifications of existing wireless LAN technique, but is most notable for its small size and low cost.
- It is envisioned that Bluetooth will be included within equipment rather than being an optional extra.
- When one Bluetooth product comes within range of another, they automatically exchange address and capability details. They can then establish 1 M bits link (upto 2 Mbps in the 2G technology) with security and error correction, to use as required.
- The protocols will handle both voice and data, with a very flexible network topography.
- This technology achieves its goal by embedding tiny, inexpensive, short-range transceivers into the electronic devices that are available today.
- The Bluetooth module can be either built into electronic devices or used as an adapter.

### 1.12.7 Functional Block Diagram

- In a PC, bluetooth can be built in as a PC card or extremely attached via the USB port. The functional block diagram of a bluetooth is as shown in Fig. 1.7.

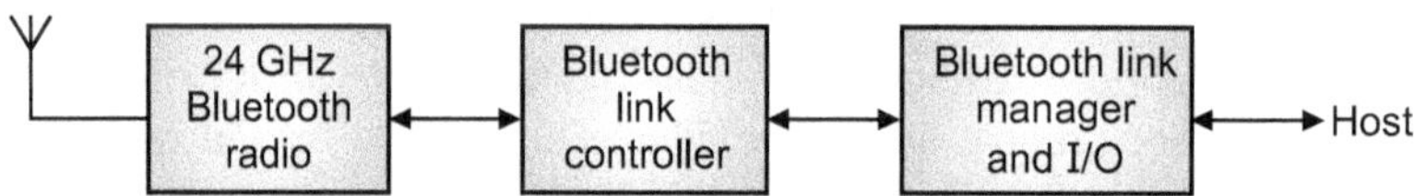


**Fig. 1.7 : Functional block diagram of bluetooth**

### 1.12.8 Principle of Operation

- Each device has a unique 45-bit address from the IEEE 802 standard.
- Connections can be point-to-point or multipoint. The maximum range of 10 meters can be extended to 100 meters by increasing the power.
- Bluetooth devices are protected from radio interference by changing their frequencies arbitrarily upto a maximum of 1600 times a second, a technique known as **frequency hopping.**
- They also use three different but complementary error connection schemes.
- Built-in encryption and verification is provided. Moreover, bluetooth devices won't drain precious battery life.
- The bluetooth specification targets power consumption of the device from a **hold** mode consuming 30 μA to the active transmitting range of 8 to 30 mA or less than 0.1 W.
- The radio chip consumes only 0.3 mA in standby mode which is less than 3% of the power used by a standard mobile phone.
- The chips also have excellent power saving features as they will automatically shift to low-power mode as soon as traffic volume lessens or stops.
- Designed to operate in a noisy RF environment, bluetooth radio uses a fast acknowledgement and frequency hopping scheme to make the link robust.
- Bluetooth radio modules avoid interference from other signals by hopping to a new frequency after transmitting or receiving a packet.
- The encoding is optimized for an uncoordinated environment. Bluetooth guarantees security at the bit level.
- The bluetooth will not interfere or cause harm to public or private telecommunications network.
- The bluetooth baseband protocol is a combination of circuit and packet switching. Slots can be reserved for synchronous packets. Each packet is transmitted in a different hop frequency.

- Bluetooth can support an asynchronous data channel upto three simultaneous synchronous voice channels or a channel, which simultaneously support asynchronous data and synchronous voice.
- It is thus possible to transfer the data asynchronously, while at the same time talking synchronously at the same time.
- Each voice channel supports 64 kbps synchronous (voice) link. The synchronous channel can support an asymmetric link of maximally 721 kbps in either direction while permitting 57.6 kbps in the return direction or a 432.6 kbps symmetric link.

### 1.12.9 Advantages

1. It is economical to implement by companies.
2. It has wireless synchronization. No need to carry connection cable while travelling.
3. It is universally accepted technology.
4. It's connectivity is automatic and hence does not need professionals.
5. It is upgradable and has backward compability with older versions.
6. It connects devices to each other irrespective of their models, as it has standard protocols.
7. It has instant Personal Area Network (PAN) consisting of upto seven bluetooth devices within a range of upto 30 feet.
8. It has faster data and voice communications sharing.
9. It simplifies the audio and communication issues arsing while driving, talking and listing music on your cell phone.
10. It reduces the battery power consumption or electrical power.
11. It avoids interference from other wireless devices with the usage of technique of frequing hopping.
12. It is best alternative to data transfer.

### 1.12.10 Disadvantages

1. The better use can be severely increased if bluetooth is enabled on cell phone while you play music.
2. If you are pushing so much data access, it is going to be slow and not to the same standard Wi-Fi.
3. As it is omni-directional, it can have problem, when it is trying to discover a recipient device like headsets, speakers, phones etc.
4. A bluetooth only offers 1 mbps, infrared is yet to be dispensed completely to have very fast rates for data transfer because infra-red needs data rates of upto 4 mbps.
5. The greater range and RF of bluetooth makes it much more open to interception and attack.

### 1.12.11 Limitations

1. Drain on battery power for smart phones.
2. Fairly limited range.
3. Difficult process.

### 1.12.12 Frequency Band

- Bluetooth operates at frequencies between 2402 and 2480 MHz or 2400 and 2483.5 MHz, including guard bands of 2 MHz wide at the bottom and 3.5 MHz wide at the top. This is the short range radio frequency (RF) band ranging from 2.4 GHz.

### 1.12.13 Specifications

1. **Core Specifications :**
   - The core specifications define the bluetooth protocol stack and the requirements for testing and qualification of bluetooth-based products.
   - The core specifications consist of five layers as under (i) Radio, (ii) Baseband layer, (iii) Link manager protocol (LMP) (iv) Logical link control and adoption protocol (LLCAP) (v) Service discovery protocol (SDP).
2. **Profile Specifications :**
   - Profile specifications define usage models that provide detailed information about how to use the bluetooth protocol for various types of applications.

### 1.12.14 Applications

1. Most of the peripheral devices such as mouse, keyboard, printer, speakers, etc. are connected to the PC cordlessly.
2. It can be used to allow one headset to be used with myriad devices, including telephones, portable computers, stereos etc.
3. It can automatically carry out certain tasks on behalf of the user without user intervention or awareness.
4. Exchanging of the multimedia data like songs, video, pictures etc. can be transferred among devices.

### 1.12.15 IEEE 802.15.1 Bluetooth Protocol

- IEEE 802.15 is a working group of Institute of Electrical and Electronics Engineers (IEEE).
- IEEE 802 standards committee which specifies Wireless Personal Area Network (WPAN) standards.
- There are 10 major areas of development, not all of which are active.
- The number of task groups in IEEE 802.15 varies based on the number of active projects.
- The task group one is based on bluetooth technology. It defines physical layer (PHL) and media access control (MAC) specifications for wireless connecting with fixed, portable and moving devices within or entering personal operating space. These standards were issued in 2202 and 2005.

## 1.13 ZIG BEE

### 1.13.1 Brief History

- *ZigBee - style* self-organizing *ad-hoc digital radio networks* were conceived in the 1990s. The IEEE 802-15.4-2003 ZigBee specification was ratified on December 4, 2004. The ZigBee Alliance announced availability of specification, 1.0 on June 13, 2005, known as the *ZigBee - 2004 specification.*
- In September 2006, the ZigBee 2006 specification was announced, obsolating the 2004 stack. The 2006 specification replaces the message and *key value pair* structure used in 2004 stack with a *cluster library.* The library is a set of standardized commands, organized under group known as *clusters* with names such as *smart energy*, Home Automation, and ZigBee light link.
- In January 2017, ZigBee Alliance renamed the library to *Dotdot* and announced it as a new protocol to be represented by an *emiticon.* They also announced that it will now additionally run over other network types using *Internet Protocol* (IP) and will interconnect with other standards, such as *Thrad.* Since, it is unveiling, Dot dot has functioned as the default application layer for almost all ZigBee devices.
- ZigBee Pro is also known as *ZigBee 2007,* was finalized in 2007. A ZigBee Pro device may join and operate on a legacy ZigBee network and vice-versa. Due to differences in routing options, ZigBee Pro devices must become non-routing ZigBee End Devices (ZEDs) on a legacy ZigBee network, and legacy ZigBee devices must become ZEDs or a ZigBee Pro network. It operates not only on 2.4 GHz band, but also on sub GHz band.

### 1.13.2 Introduction

- The technology defined by the ZigBee specification is intended to be simple and less expensive them other *Wireless Personal Area Networks (WPANs),* such as *Bluetooth* or more general wireless networking such as *Wi-Fi.* Applications include wireless light switches, home energy monitors, traffic management systems, and other consumer and industrial equipment that requires short-range low-rate wireless data transfer.
- It's low power consumption limits transmission distances as 10 to 100 meters *line of sight,* depending on power output and environmental characteristics. ZigBee devices can transmit data over long distances by passing data through a *mesh network* if intermediate devices to reach more distances. ZigBee is typically used in low data rate applications that require long battery life and secure networking.

- ZigBee has a defined rate of 250 kbits per second (Kbps), best suited for intermittent data transmission from a sensor or input device. ZigBee was concerned in 1998, standardized in 2003, and revised in 2006. The name refers to the *waggle device* of honey bees after their return to the *believe*.

### 1.13.3 Definition

- ZigBee is a low-cost, low-power, wireless mesh network standard targeted at battery powered devices in wireless control and monitoring applications.
- **ZigBee i.e. an IEEE 802.15.4 based specification** for a suite of high-level communication protocols used to **create personal area networks with small, low power digital radios,** such as for home automation, medical device data collection, and other low-bandwidth needs, designed for small scale projects, which need wireless connection.
- **ZigBee is an open global standard for wireless technology deployed designed to use low-power digital radio signal for personal area networks** (PANs).
- ZigBee operates on the IEEE 802.15.4 specification and is used to create networks that requires a low-data transfer rate, energy efficiency and secure networking.

### 1.13.4 Architecture

- ZigBee delivers low-battery communication. ZigBee chips are typically integrated with radios and with microcontrollers. It operates in the industrial, scientific and medical (ISM) radio bands 2.4 GHz in most jurisdictions world wide through some devices use 784 MHz in China, 863 MHz in Europe and 915 MHz in the U.S. and Australia. However, even those regions and countries still use 2.4 GHz for most commercial ZigBee devices for home use.
- ZigBee builds on the *physical layer* and *media access control* defined in IEEE standard 802.15.4 for low-rate wireless personal area networks (WPANs). The specification includes four additional key components : network layer, application layer, ZigBee Device Objects (ZDOs) and manufacturing defined application objects. ZDOs are responsible for some tasks, including keeping track of device roles, managing requests to join a network, as well as device discovery and security.
- The ZigBee *network layer* natively support both star and *tree* networks, and generic *mesh* networking.

### 1.13.5 Description

- In the communication world, there are numerous high data rate communication standards that are available but none of these meet the sensors and control devices communication standards. ZigBee system structure consists of three different types of devices such as ZigBee coordinator, Router and End device.
- Every ZigBee network must consist of at least one coordinator, which acts as a root and bridge of the network. The coordinator is responsible for handling and storing the information, while performing, receiving and transmitting data operations.
- ZigBee routers act as intermediary devices that permit data to pass to and fro through them to other devices. End devices have limited functionality to communicate with the patent modes, such as the battery power is rolled. The number of routers, coordinators and end devices depend on the type of networks, such as star, tree and mesh.
- ZigBee protocol architecture consists of a stack of networks. Various layers where IEEE 802.15.4 is defined by physical and MAC layers, while the protocol is completed by accumulating ZigBee's own network and application layers.
- The physical layer does modulation and demodulation operations upon transmitting and receiving signals respectively. This layers frequency, data rate and number of channels.

- The network layer takes care of all network related operations such as network set-up and device connection and disconnection to network, routing, operations such as network set-up end device connection and disconnection to network, routing, device, configurations etc.
- The architectural structure of a ZigBee protocol is as shown in Fig. 1.8.

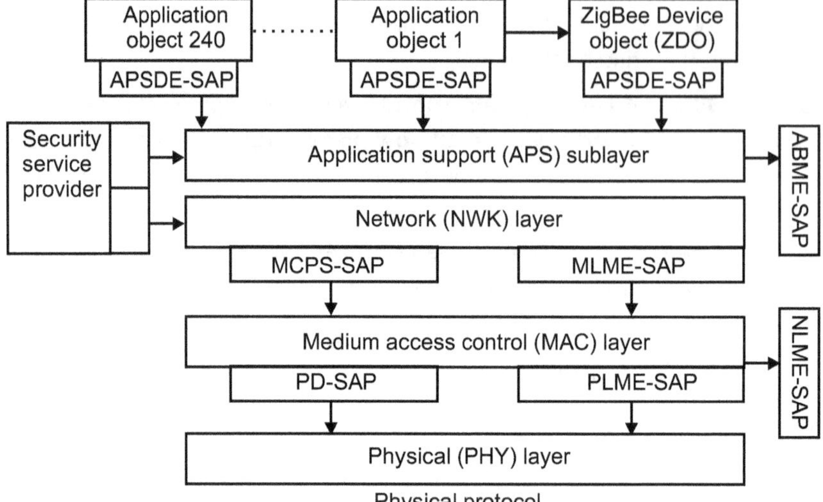


**Fig. 1.8 : ZigBee protocol architecture**

- MAC layer is responsible for reliable transmission of data by accessing different networks with the carrier sense multiple access collision avoidance (CSMA). This also transmits the Beacon frames for synchronizing communication.
- Application support sub-layer enables the services necessary for ZigBee device object and application objects to interface with the network layers for data managing services. This layer is responsible for matching two devices, according to their services and needs.
- The application framework provides two types of data services as key value pair and generic message services. Generic message is a developer defined structure, whereas the key value pair is used for getting attributes within the application objects and APS layer in ZigBee devices. It is responsible for detecting, initiating and binding other devices to the network.

## 1.13.6 Digital Cellular System

- Digital cellular systems are the cellular systems, that use the digital communication techniques, like modulation, transmission formation and demodulation and so on. The characteristics of these systems are:
  1. These after an effective data transmission compared to the conventional analog cellular systems. These systems employ the packet switched communication technique, which is faster than the circuit switching technique.
  2. These systems employ powerful error detection and correction.
  3. These systems due also provide the security on transmitting data through **encryption** and **decryption technique authentication**.
  4. These systems also require very less transmit power, the property increases the battery life in portable mobile units.
  5. The range of services provided by the digital cellular system is quite high as compared to that provided by the analog cellular system.
  6. The speed of services provided by digital systems is quite high and thus, they support high capacity data transfer.
  7. These systems employ IDMA technique for digital communication.

- The newer cellular systems, such as GSM, CDMA and PCs are digital cellular systems. In 1992 the first digital cellular system, GSM was developed in Germany. GSM is a European standard system. In United States, NA - TDMA system and CDMA systems have been developed. A Japanese system, PDC (Personal Digital Cellular) was deployed in Osaka in June 1994.

### 1.13.7 Topologies

1. Star
2. Mesh
3. Cluster trees.

### 1.13.8 Explanation

- ZigBee is a cost and energy efficient wireless network standard. It employs mesh network topology, allowing it to provide high reliability and reasonable range.
- One of ZigBee's defining features is the secure communications, it is able to provide. This is accomplished through the use of 128 bit cryptographic keys. This system is based on symmetric keys, which means that both the recipient and originator of a transaction need share the same key.
- These keys are either pre-installed, transported by true-center designated within the network or established between the trust center and device without being transported. Security in a personal area network is most crucial, when ZigBee is used in corporate or manufacturing networks.

### 1.13.9 Device Types

1. ZigBee coordinator (ZC)
2. ZigBee router (ZR)
3. ZigBee end device (ZED).

### 1.13.10 Maximum Range

- The low power consumption of ZigBee limits the transmission distance to 10 - 100 meters line-of-sight, depending on power output and environmental characteristics. ZigBee devices can transmit data over long distances by passing data through a mesh network of intermediate device to reach more distance ones.

### 1.13.11 Application Areas

1. Home automation
2. Wireless sensor networks
3. Industrial control systems
4. Embedded sensing
5. Medical data collection
6. Smoke and intruder warning
7. Building automation
8. Remote wireless microphone configuration.

### 1.13.12 Applications

1. Industrial automation
2. Home automation
3. Smart metering
4. Smart grid monitoring
5. Photo credits.

### Practice Questions

1. What is mobile communication ?
2. State advantages of mobile communication.
3. State important applications of mobile communication.
4. What is a cellular system ?
5. Draw schematic diagram of a basic cellular system and describe it in brief.

6. Explain the operation of a cellular system,
7. Explain the planning of a cellular system.
8. Compare analog and digital cellular systems.
9. What is wireless communication ?
10. Give various examples of wireless systems.
11. Write a short note on following wireless systems :
    (a) Paging system,
    (b) Cordless telephone system,
    (c) Cellular telephone system.
12. What is Bluetooth technology ?
13. What are the applications of Bluetooth ?
14. List the key features of Bluetooth.
15. State basic principle of Bluetooth.
16. Draw functional block diagram of Bluetooth and explain it.
17. State advantages and limitations of Bluetooth.
18. What is ZigBee ?
19. Describe ZigBee in short.
20. What is Bluetooth technology ?
21. What are the applications of Bluetooth ?
22. Draw functional block diagram of Bluetooth.
23. Explain basic principle of Bluetooth.
24. Explain principle of operation of Bluetooth.
25. What are advantages and applications of Bluetooth ?

✍✍✍

Chapter 2

# ELEMENTS OF CELLULAR RADIO SYSTEMS DESIGN

**Syllabus**

General description of the problem, Concept of frequency reuse channels, Co-channel interference, Reduction factor, Desired C/I from a normal case in an omni-directional antenna system, Cell splitting, Considerations of the components of cellular systems.

## 2.1 MOBILE RADIO TELEPHONE SYSTEM

### 2.1.1 Introduction

- In 1946, in St. Louis (USA), the *Mobile Telephone Service* was first introduced. The main aim of the early mobile radio system was to provide the coverage to a large area with the help of a single high power transmitter having an antenna mounted on a tall tower.
- But this mobile radio system suffers from number of problems. It has limited radio coverage area and limited radio spectrum. Therefore, this type of mobile radio system is not used in practice. Hence, it becomes necessary to restructure the mobile radio telephone system to enhance the user capacity, also the proper spectrum allocation in proportional with increasing demand and also the larger coverage area.
- In 1964, improved mobile telephone service was introduced with additional channels and more automatic handling of calls to the Public Switched Telephone Network (PSTN).
- Even the addition of radio channels in three bands was insufficient to meet the demand for vehicle mounted mobile radio systems. The mobile radio-telephone services before the first generation (1G) of cellular telephones, these systems are sometimes referred as *pre-cellular* for sometimes zero generation systems.
- Technologies used in pre-cellular systems included the push to talk (PTT or manual), Mobile Telephone System (MPS), Improved Mobile Telephone Service (IMTS), and Advanced Mobile Telephone System (AMTS) systems. These mobile telephones were usually mounted in cars or trucks through briefcase models were also made. Typically, the transceiver (transmitter-receiver) was mounted in the vehicle such as truck and attached to the *'head'* (dial, display, and bandsets) mounted near the driver seat.

### 2.1.2 Limitations

1. **Limited service capability :** There is no guarantee that a call can be completed without a hand-off.
2. **Poor service performance :** The number of active users is limited to the number of channels assigned to a particular frequency zone.
3. **Inefficient frequency spectrum utilization :** The maximum number of customers that could be served by one channel at a busy hour is limited due to inefficient frequency spectrum.

### 2.1.3 Disadvantages

1. The call is dropped, when moved from one zone to other.
2. The user has to reinitiate the call, when moved to other zone.
3. The number of active users is limited to the number of channels assigned to a particular frequency zone.
4. The large number of subscribers create a high blocking probability during busy hours.
5. It has inefficient frequency spectrum utilization.
6. It is not possible to reuse the frequency spectrum.
7. It is not possible to have the proper allocation of frequency spectrum in proportion with the increasing demand.

## 2.2 CELLULAR CONCEPTS

### 2.2.1 Introduction

- In December 1947, *Douglas A. Ring and W. Ras Young,* engineers of Bell Laboratories proposed *hexagonal cells* for mobile phones in vehicles.
- *Philip T. Peter* also of Bell Laboratories proposed that the cell towers be at the corners of the hexagons rather than the centers and have directional antennas that would transmit/receive in three directions into three adjacent hexagon cells on three different frequencies. At this stage, the technology to implement these ideas did not exist, nor had the frequencies been allocated. Several years would pass before *Richard H. Frankiel* and *Joel S. Engel* of Bell Laboratories developed the electronics to achieve this in the 1960s.
- In all these early examples, a mobile phone had to stay within the coverage area serviced by one base station throughout the phone call, i.e. there was no continuity of service as the phones moved through several cell areas.
- The concepts of *frequency reuse* and *handoff* as well as number of other concepts that formed the basis of modern cell phone technology, were described in the 1970s.
- The 1970, *Amos E. Joel,* Jr. engineer of Bell Laboratories invented an automatic *'call hand off'* system to allow mobile phones to move through several cell areas during a single conversation without interruption. In 1969, *Amtrak* equipped computer trains along the 225 mile *New York - Washington* route with special pay phones that allowed passengers to place telephone calls while the train was moving. This system reused six frequencies in the 450 MHz band in nine sites, a procedure of the concept later applied in cellular telephones. In December 1971, AT and T submitted proposal for cellular service to the *Federal Communications Commission* (FCC). After years of hearings, the FCC approved the proposal in 1982 for Advanced Mobile Phone System (AMPS) and allocated frequencies in the 824-894 MHz band. *Analog AMPS* was eventually superseded *Digital AMPS* in 1990.
- A cellular telephone switching plan was described by *Fluhar and* Nursbaum in 1973 and a cellular telephone data signaling system was described in 1977 by *Hachenburg et al.*

### 2.2.2 Fundamental Concept of a Cell

- Cellular telephony derives its name from the portion of a geographic area into small *'cells'.* The basic concepts and theory of cellular telephone was developed by AT&T Bell Laboratories of United States in 1947, but the first tests were conducted in 1962 to explore commercial applications.
- The cellular telephone systems rely on an intelligent allocation and reuse of channels throughout a coverage region.
- With cellular concepts, each area (region) is further divided into *hexagonal* shaped cells that fit together to form a honey comb pattern. The hexagon pattern was chosen, because it provides the most efficient transmission by approximating a circular pattern, while eliminating the gaps inherently present between the adjacent circles in cellular communication.
- The number of cells per system and size of cells are decided by service providers depending upon the traffic patterns. Each *geographic* area is allocated a fixed number of voice channels. The physical area of cell varies according to user (subscriber) density and calling patterns (traffic patterns). Each cellular base station is allocated a group of radio channels to be used within a small geographic area called a Cell.

### 2.2.3 Definition

- **A small geographic area of a base station covered by cellular radio antenna is called a cell.** Thus, a cell represents the coverage area of a base station.
- ***The base station transmitting over a small geographic area is called a cell.*** Cell is defined by its physical size, and size of its population and traffic pattern.
- The number of cells per system and size of the cells are decided by service providers depending upon traffic patterns. The physical size of the cell varies according to user subscriber density and calling patterns (traffic patterns).

- ***The basic geographic unit of cellular communication system is known as a cell.***
- The large cells were called *macrocells*. Typically they have radius between 1.5 to 25 km with base station transmission power between 1W and 6W.
- The smallest cells called *microcells* typically have radius of 0.45 km or less than 0.45 km with base station transmitting power between 0.1W and 1W. Microcells are used mostly in high density area such as large cities and inside buildings.
- Sometimes cellular radio signals are very weak, then very small cells called *picocells* are used. Indoor picocells are used in the areas like underground malls. Here, indoor picocells use same frequencies as regular cells.

## 2.2.4 Shape of a Cell

- What should be the geometric shape of a cell ? While deciding the geometric shape of a cell, we have to consider such a geometric shape which covers the entire region of a radio coverage of a base station without overlap and has equal area of radio coverage.
- By considering these two factors, there are three sensible choices for the geometric shape of a cell : (1) a square, (2) an equilateral triangle, and (3) a hexagon. A cell must be designed to serve the weakest mobiles within the foot prints which are typically located at the edge of the cell.
- It might be seen natural to choose a *circle to* represent the coverage area of a base station, but adjacent circles cannot be overlaid upon a map without leaving gaps or creating overlapping regions. So the circle cannot be a suitable choice for the shape of a cell. Also an equilateral triangle does not fulfill the required design conditions of a cell. For a given distance between the center of a *polygon* and its farthest perimeter points, the hexagon has the largest area as that of a circle and an equilateral triangle.
- Thus, by using the *hexagon* geometry, the fewest member of cells can cover a geometric region of a base station and also the *hexagon* closely approximates a circular radiation pattern, which would occur for an omni-directional base station antenna and free space propagation. So, the hexagon is a best choice for the geometric shape of a cell.
- The hexagonal shape of a cell shown in Fig. 2.1 is a conceptual and a simplest model for a cell of the radio coverage for each station.
- It has been universally adopted, since the *hexagon* permits easy and manageable analysis of a cellular measurements or propagation prediction models.

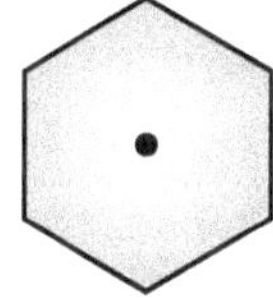

**Fig. 2.1 : A Hexagonal cell**

- In designing the hexagonal shaped cells, base station transmitters can be located as under :

  (a) in center of cell known as center excited cell, (b) at edge (edge excited cell), or

  (c) at corner (known as corner excited cell).

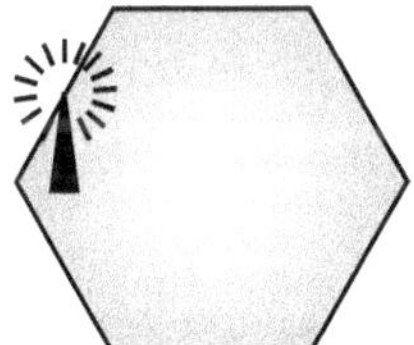

**Fig. 2.2 : Excited cells**

- Omnidirectional antennas are used in center excited cell. Sectored directional antennas are used in the edge excited and corner excited cells. Thus, instead of only one high power transmitter for large geographical area, the cellular communications implement small power multiple transmitters in hexagonal small cells.

### 2.2.5 Cell Cluster

**(a) Definition :**

- **A group of cells that uses all the available set of frequencies is known as a cluster.**

**(b) Effect of cluster size :**

- The cluster size is not fixed, but it depends on the requirements of a particular area of a cluster size. It is denoted by a letter 'n'.
- The size of various clusters is shown in Fig. 2.3.

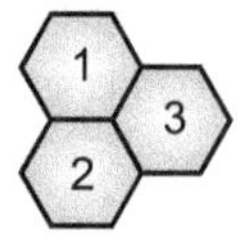


(a) n = 3

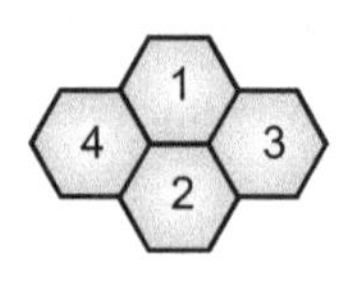


(b) n = 4

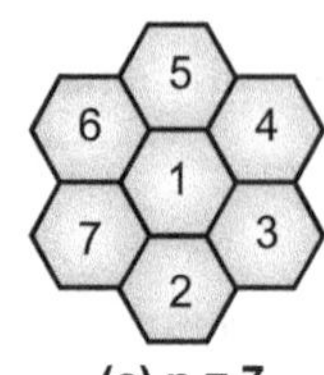


(c) n = 7

**Fig. 2.3 : Cell clusters for frequency reuse**

- Thus, Fig. 2.3 (a), (b) and (c) show the three cell cluster (n = 3), four cell cluster (n = 4) and seven cell cluster (n = 7) respectively. Note how neatly 7 hexagon shaped cells fit together. Try that with a triangle. Cluster of 4 and 12 cells are also possible but frequency reused patterns based on 7 are most efficient.

### 2.2.6 Advantages

1. It has better user capacity.
2. It has higher coverage range and area.
3. It has no frequency congestion.
4. It has made possible to introduce the concept of frequency reuse.
5. It requires only a fixed number of channels (i.e. frequency slots).
6. It has made possible to use the proper frequency spectrum allocation with the increasing demand.
7. It can use low power transmitters.
8. It requires no major technological changes.
9. It has made possible to manufacture the every mobile handset within a country or continent with the same set of channels. So many mobile handset can be used anywhere.

## 2.3 CELLULAR RADIO TELEPHONE SYSTEMS

### 2.3.1 Need

- The conventional mobile telephone systems suffer from many demerits and limitations, namely; (1) Limited service capability, (2) Poor service performance, and (3) Inefficient frequency spectrum utilization.
- Therefore, to overcome the demerits and limitations of conventional mobile telephone systems, and also to achieve the objectives of (1) high capacity of user, (2) limited frequency spectrum, and (3) large coverage area, there is a necessity of cellular mobile telephone systems.

### 2.3.2 Introduction

- By 1930, telephone customers in the United States could by a call to a passenger on a line in the Atlantic Ocean. In areas with *Marine VHF radio* and a shore station, it is still possible to arrange a call from the public telephone network to a ship, still using natural call set-up and the services, if a human marine radio operates.
- However, it was the 1940s onwards that saw the seeds of technological development, which would eventually produce the mobile phone that we know today.
- Motorola developed a backpacked two-way radio, the *Walkie-Talkie* and a large hand-held two-way radio for the United State Military.
- The battery powered *Handle-Talkie* (HT) was about the size of a man's forearm. Prior to 1973, cellular mobile phone technology was limited to phones installed in car and other vehicles. On April, 3, 1973, *Martin Cooper,* Motorola researcher and executive, made the first analog mobile phone call using a heavy prototype model. He called *Dr. Joel S. Engel* of Bell Laboratories. There was a long race between *Motorola* and *Bell Laboratories* to produce the first portable mobile phone.

- *Martin Cooper* is the first inventor named in *'Radio Telephone System'* filed on October 17, 1973 with *US patent office* and later issued as US patent 3096, 166. *John F. Mitchell*, Motorola's chief of portable communication product (and cooper's boss) was also named on the patent. He successfully pushed Motorola to develop wireless communication products that would be small enough to use anywhere and participate in the design of the cellular phone.

### 2.3.3 Concept

- In 1970, Bell Laboratories of United States proposed to build the high capacity cellular telephone system and then introduced in the early 1970s. One of the most successful initial implementations of the cellular concept was the Advanced Mobile Phone System (AMPS) which has been available in the United States, since 1983.
- **A cellular radio system is generally characterized as a high capacity land mobile system in which available frequency spectrum is partitioned into the discrete channels which are assigned in groups to geographic cells covering a cellular Geographic Service Area (GSA). The discrete channels are capable of being reused in different cells within the service area.**
- The cellular (radio) phone is a wireless (radio) communication system just like a cordless phone. The cellular phone is also known as cell phone. In cell phone, the distance is not restricted to within home but one can travel in the city or even outside the city without interruption in communication.

### 2.3.4 Basic Principle

- The basic principle of a cellular (radio) system is to divide a large geographic service area into cells with diameters from 2 to 50 km each of which is allocated a number of radio frequency (RF) channels. The transmitters in each adjacent cell operate on different frequencies to avoid the interference.
- Since, however, transmit power and antenna height in each cell are relatively low, the cells that are sufficiently far apart can reuse the same set of frequencies without causing co-channel interference. The theoretical coverage range and capacity of a cellular system are therefore unlimited.
- As the demand for cellular mobile service grows, the additional cells can be added, and as traffic demand grows in a given area, the cells can be splitted to accommodate the additional traffic. A cellular (radio) system provide the capability to hand-off calls in progress, as the mobile terminal (or user) moves between cells.

### 2.3.5 Description

- The basic structure of a mobile phone network along with the public switched telephone networks is shown in Fig. 2.4. It has hexagonal shaped cells as shown in Fig. 2.1. Each cell has a base station (BS) situated at the center. The task of the base stations is to act as an interface between the mobile phone and the cellular radio system.
- The base stations of all cells are connected to the switched center. The interface between the switched station and the base stations is a bidirectional that means the exchange of information between the switched center and the base station is a two way. The communication area of the mobile communication is divided into hexagonal cells. Therefore, the system is named as cellular radio system.
- The switching center acts at the interface between the PSTN and the calls. In addition to that it performs the supervision and control operations in the mobile communication system. Due to this kind of a system layout, the communication can take place between two mobile subscribers or station of that cell. Thus, the service provided to a mobile subscriber is continuous without any break.

## 2.4 CELLULAR SYSTEM

### 2.4.1 Architecture

- Fig. 2.4 shows a typical basic cellular system architecture. It consists of a Base Stations (BS), a Home Location Register (HLR), a Mobile Switching Centre (MSC), a Visiting Location Register (VLR) and a Public Switched Telephone Network (PSTN).

## 2.4.2 Operation

- Consider the architecture of a cellular system as shown in Fig. 2.4.
- The coverage area of a cellular system is partitioned into a number of smaller areas i.e. cells with each cell served by a Base Station (BS) for radio coverage.
- The base stations are covered through fixed links to a Mobile Switching Center (MSC), which is a local switching exchange (MTSO) with additional features to handle mobility management requirements of a cellular system.
- To accommodate the dynamic nature of terminal location information and subscriptional data, the MTSO interacts with same form of database that maintains the subscriber data and location information.

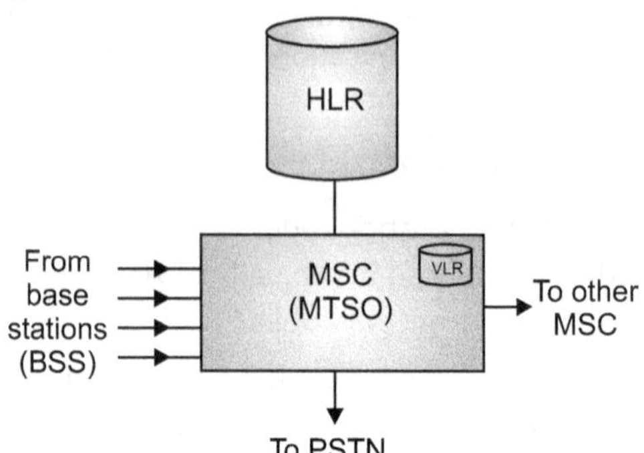


**Fig. 2.4 : Cellular system architecture**

- The MTSO also interconnects with the PSTN, because the majority of calls in a cellular mobile system either originate from or terminate at fixed network terminals. Based on the frequency spectrum made available by the licencing authority and the cellular standard in use, the cellular system is able to define a number of radio channels for use across it serving area.
- The available radio channels are then partitioned into groups of channels and these groups of channels are allocated to individual cells forming the entire serving area. The individual channels or a particular groups of channels can be reused in the cells that are located far enough apart.
- A key feature of radio system planning activity consists in designing cell sizes, assigning their locations and allocating radio channels to individual cells.
- In each cell, one radio channel is set aside for carrying signaling information between the networks, i.e. the base station and the mobile station in that cell. The signaling channel is used in the mobile to base station direction to carry signals for location updating, mobile originated call set up and responses to incoming call set up messages e.g., paging response. In the reverse direction, i.e. base station to mobile, the signaling channel carries message related to operating parameters e.g., location area identity, cell identity, call set up (e.g., paging) and location updating.
- The Home Location Register (HLR) represents a centralized database that has the permanent data fill about the mobile subscribers in a large service area (generally one per GSM network operator). It is kept updated with the current location of all its mobile subscribers, including those who may have removed to another network operator within or outside the country. Besides the up-to-date location information for each subscribers, which is dynamic, the HLR maintains the subscriber data on a permanent basis.
- The Mobile Switching Centre (MSC) for Groups Special Mobile (GSM) can be viewed as a local ISDN switch with additional capabilities to support mobility management functions like terminal registration, location updating and hand-off. Further, unlike a local switch in a fixed network, the MSC does not contain the mobile subscriber parameters. The Visiting Location Register (VLR) represents a temporary data, data store and generally there is one VLR per MSC.
- This register contains information about the mobile subscribers who are currently in the service area converted by MSC/VLR. The VLR also contains information about locally activated features such as call forward on busy. The Base Station System (BSS) comprises a Base Station Controller (BSC) and one or more subtending Base Transceiver Station (BTS). The BSS is responsible for all functions related to the radio resource (channel) management.
- This includes the management to radio channel configuration with respect to use a speech, data or signaling channels, allocation and release of channels for call set up and release control of frequency hopping and transmit power at the mobile station.

## 2.5 CELLULAR TELEPHONE SYSTEM

### 2.5.1 Introduction

- Cellular telephone, sometimes called *mobile telephone*, is a type of short-wave analog or digital telecommunication in which a subscriber has a wireless connection from a mobile phone to a relatively nearby transmitter. As the cellular telephone user moves from one cell or area of a coverage to another, the telephone is effectively passed on the local cell transmitter.
- Cellular telephone systems are designed to provide two-way voice communication at vehicle speeds with regional or national coverage. A cellular telephone system provides a wireless connection to the PSTN for any user location within the radio range of the system.

### 2.5.2 Schematic Diagram

- Fig. 2.5 shows the schematic diagram of a basic cellular telephone system. It gives the interconnection of elements of cellular telephone system.
- It is also called AMPS. It is a method to provide high quality mobile service for more subscribers at an affordable cost and to provide more freedom to roamers. A basic cellular telephone system consists of three subsystems : (1) mobile unit, (2) a cell site and (3) a Mobile Telephone Switching Office (MTSO), with connections to link these three subsystems.

### 2.5.3 Description

The function of each unit of a basic cellular radio system is described below :

1. **Mobile Handset (Unit) :**

- The mobile telephone handset is a compact portable handset unit. It is an extremely complex piece of digital engineering made possible by advanced ICs and advancements in telecommunication technology. The mobile handset consists of control unit, trans-receiver and antenna system.
- Functionally, the mobile handset itself can be divided into three main sections namely terminal adaptations, radio modem and RF unit.
- The termination adoptions comprise the human user interface, elements like microphone, speaker, display and keyboard and interfaces to other equipments such as a PC or PCMCIA data card.
- The radio modem is a Digital Signal Processing (DSP) unit which handles the conversion of data or speech signals into digital form.
- It interfaces between the terminal adaptations and the RF unit.
- The RF unit handles the actual wireless communication by receiving and transmitting radio signals to the cellular network. It has also a Subscriber Identity Module (SIM).
- The SIM is a credit card sized plastic module which fits into the mobile phone. This SIM is smart card and contains all the subscribers related information, like your identification number and preferences.
- It can also be used for storing 10 digits phone numbers and messages.

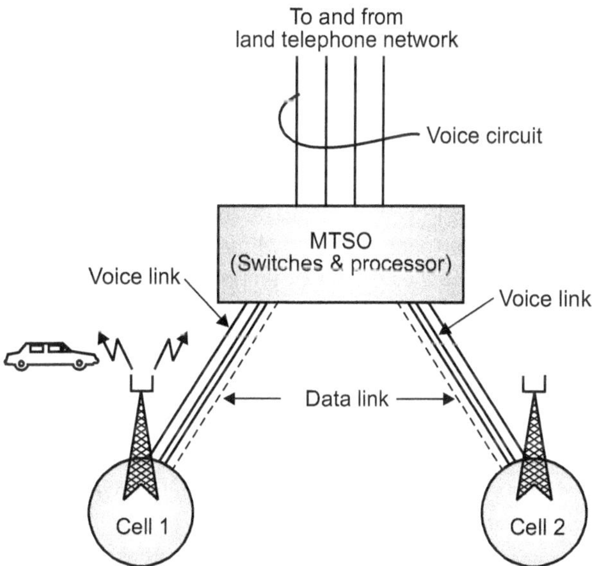


**Fig. 2.5 : Basic cellular telephone system**

- It also enables charges to be automatically billed to the card holder regardless of who owns the mobile unit. This means that if your mobile handset is out of order, you can always use any other handset with your SIM card without affecting the other person's billing.

- The power level can be controlled by the Base Station (BS) so that strength of the signal is minimum required which helps in avoiding interferences. These mobile handsets can be used for wireless data connections and wireless internet access. They can also be used as way digital messaging units.
- Whenever, mobile handset is turned on but not in use, the control unit monitors the data being transmitted on a set up channel selected from among the several standard set up frequencies on the basis of signal strength. If the signal strength becomes marginal as the mobile unit approaches a cell boundary, the mobile control finds a set up channel with a stronger signal.
- When you dial a number on the keypad of the mobile handset, the handset transmits the digits, through the built in radio transreceiver to a nearby radio base station (also called cell). A group of transreceiver stations are controlled by a base station controller, which in turn is connected to Mobile Switching Centre (MSC). The MSC, in turn, is linked to the other cellular and fixed line networks.

**2. Cell Sites :**

- Each cellular service area is divided into small regions called *cell* for making the tracking possible. Each cell is equipped to switch, transmit and receive calls to and from any module unit located in the cell. Each cell transmitter and receiver operates on a given channel. Each channel is used for many simultaneous conversations in cells which are not adjacent to one another, but are for enough apart to avoid excessive interference.
- Thus, a system with a relatively small numbers of subscribers can use larger cells and as demand grows, the cells are divided into smaller ones. Each cell has all stations called as cell sites. The cell sites from the radio link between individual cellular telephones and the telephone system. Each cell station is equipped with a transmitter and receiver coupled to an array of antennas. The cell site consists of a control unit, radio cabinet, antenna power plant and data terminals. It is located where it will operate most effectively in the radio environment.
- In urban areas, it may be found on a top hill building. In suburban or rural areas, it may be located on large hills or mountains, wherever the best radio coverage can be obtained. *It provides interference between MTSO and the mobile unit.*

**3. MTSO :**

**(a) Concept :**

- Mobile Telephone Switching Office (MTSO) is the heart of the cellular mobile system. It consists of switching office, cellular processor, cellular switch and a computer. The transmitted power levels at both the cell sites and the mobile units by the MTSO (or the cell site) and not the mobile units.
- There can be only limited power control by the cell sites as a result of system limitations. The mobile unit can lower the power level but cannot arbitrarily increase in it. This is because the MTSO is capable of monitoring the performance of the whole system and can increase or decrease the transmitted power level of those mobile units to render optimum performance.
- The MTSO will optimize performance for any particular mobile unit unless a special arrangement is made. It helps to interface with telephone company zone offices. It controls call processing and does billing activities. Its processor is used for central coordination and cellular administration. It is a computerised centre that is responsible for connecting calls as well as recording call information and billing.
- The cellular switch used may be analog type or digital type. It helps to connect the one mobile subscriber to other mobile subscriber and also nation wide telephone network. It uses data links for control and supervision between processor and the switch and between the cell site and the processor. The radio link carries the voice and signaling between the mobile unit and the cell site.
- The MTSO uses a substantial amount of additional digital equipments programmed for cellular control. It not only connects the telephone network, but also records call information for billing purposes, the MTSO is linked to the cell sites by a group of voice trunks for conversations, together with one or more data links for signaling and control. The MTSO controls not only the cells sites via radio commands, but also many functions of the mobile units.

**(b) Functions of MTSO :**

The important functions performed by the MTSO are as under :

1. It controls channel assignment.
2. It controls call processing and does billing activities.
3. It controls call set-up and call termination.
4. It includes signaling, switching, supervision and allocating RF channels.
5. It controls all the cells and also helps to provide the interface between each cell and the telephone company zone officers.
6. As the vehicle moves from one cell to the next, it automatically switches from one cell to the next without any interruption in communication.
7. The computer at the MTSO causes the transmission from the vehicle to be switched from the weaker cell to the stronger cell with a very short time.
8. The computerized MTSO centre automatically connects calls as well as records the call information and does billing work.

**(c) Advantages of MTSO :**

The important advantages of MTSO over the older MTS system are as under :

1. It operates at much higher radio frequency so that it has wide frequency spectrum and large number of channels.
2. It allows to introduce the concept of frequency reuse.
3. It helps to accommodate the cells of different size.

**4. Connections :**

- The above mentioned three sub-systems are connected by the radio and high speed data links.
- Only one channel will be used by each mobile unit for its communication at a time.
- The channel may be any one in entire band assigned by the serving area, with each site having multichannel capabilities that can contact simultaneously to many mobile units.
- The channel is not fixed.

## 2.5.4 Operation

- The operation of cellular telephone system is described below from a customer's perception without considering the design parameters. This operation is divided into four parts and a hand-off procedure.

**1. Mobile Unit Initialization :**

- When a user sitting in a car activates the receiver of the mobile unit, the receiver scans 21 set up channels which are designated among the 416 channels. It then selects the strongest and locks ON for a certain time. Since each site is assigned a different set up channel, locking onto the strongest set up channel usually means selecting the nearest cell site. This self location scheme is used in the idle stage and is user independent.
- It has a great advance because it eliminates the load on the transmission at the cell site for locating the mobile unit. The disadvantage of the self-location scheme is that no location information of idle mobile units appears at each cell site. Therefore when the call initiates from the landline to a mobile unit, the paging process is longer.
- Since a large percentage of calls originates at the mobile unit, the use of self-location schemes is justified. After 60 seconds, the self-location procedure is repeated. In the future, when landline originated calls increase, a feature called *registration* can be used.

**2. Mobile Originated Call :**

- The user places the called number into an originating register in the mobile unit, checks to see that the number is correct and pushes the *send* button. A request for service is send on a selected set up channel obtained from a self-location scheme.
- The cell site receives it, and in directional cell sites, selects the best directive antenna for the voice channel to use. At the same time the cell site sends the request to the Mobile Telephone Switching Office (MTSO) via a high-speed data link.

- The MTSO selects an appropriate voice channel for the call and the cell site acts on it through the best directive antenna to link the mobile unit. The MTSO also connects the wire-line party through the telephone company zone office.

3. **Network Originated Call :**

- A land line party dials a mobile unit number. The telephone company zone office recognizes that the number is mobile and forwards the call to the MTSO.
- The MTSO sends a paging message to certain cell sites based on the mobile unit number and the search algorithm. Each cell site transmits the *page* on its own set up channel, locks on to it, and responds to the call site. The mobile unit also follows the instruction to tune to an assigned voice channel and initiate user alert.

4. **Call Termination :**

- When the mobile user turns off the transmitter, a particular signal (signaling tone) transmits to the cell site and both sites free the voice channel. The mobile unit resumes monitoring pages through the strongest set up channel.

5. **Hand-off Procedure :**

- During the call, two parties are on a voice channel. When the mobile unit moves out of the coverage area of a particular cell site, the reception becomes weak. The present cell site requests a hand-off.
- The system switches the cell to a new frequency channel in a new cell site without either interrupting the call or alerting the user. The call continues as long as the user is talking. The user does not notice the hand-off occurrences. Hand-off was first used by the AMPS system, then renamed as hand over by the European system because the different meanings in English and American English.

### 2.5.5 Performance Criteria

- There are three categories for specifying the performance criteria used for mobile communication.
  1. Voice quality,
  2. Service quality,
  3. Special features.

## 2.6 FREQUENCY REUSE CONCEPT

### 2.6.1 Definition

- To ensure that the mutual interference between users remains below a harmful level, adjacent cells are of different frequencies. However, in cells that are separated further away, frequencies can be reused.
- **Frequency reuse is the process of using the same radio frequencies on radio transmitter areas within a geographical area that are separated by sufficient distance to cause minimal interference with each other.**
- *The design process of selecting and allocating RF channel groups for all of the cellular base stations within a cellular system is called frequency reuse.*
- Cellular frequency reuse is the principles behind the working of modern day mobile communications systems. In this a group of cells referred to as a cluster is allocated a particular, frequency, and all cells in the cluster share the same frequency.
- Frequency reuse is the process, in which the same set of frequencies (channel) is allocated more than on cell provided the calls are properly separated by suffice device. Thus, cells using same set of radio channels can avoid mutual interference, if they are properly separated.

### 2.6.2 Significance

- The significance of frequency reuse is to serve the greatest number of customers and users without any co-channel interference with a specified system quality. It uses the selected and allocated group of RF channels for all of the cellular base stations within a cellular system. The frequency reuse covers the different geographic areas that are physically separated from each other.

### 2.6.3 Need

- In cellular mobile (radio) systems, it is very much important to serve the greatest number of users (customers) without any co-channel interference, with a specified system quality. The cellular system rely on an intelligent allocation of frequency channel that will keep the interference within the tolerable limits. Hence any group of frequency channels cannot be used. Therefore the same frequency channels has to be used simultaneously for different conversations in the same geographic area called cell. So there is a need to introduce the concept of frequency reuse in the cellular mobile radio systems.

### 2.6.4 Concept

- The base stations in adjacent cells are assigned RF channel groups which contain completely different channels than neighbouring cells. The base station antennas are designed to achieve the desired coverage within the particular cell. By limiting the coverage area to within the boundaries of a cell, the same group of RF channels may be used to cover different cells that are separated from one another by distances large enough to keep the co-channel interference levels within tolerable limits.
- Thus the frequency reuse means using the same frequency channel simultaneously for different conversations in the same general geographic area (cell).
- **Frequency reuse refers to the use of the radio channels operating on the same center frequency, to cover different geographic areas that are physically separated from each other by sufficient distance so that cochannel interference is not objectionable.**
- It is very important concept of the cellular mobile radio system. The users located in different geographical areas i.e. different cells can use the same frequency simultaneously. Frequency reuse is an important concept because in this a single transmitter of higher power need not be used to cover the entire geographical area. But many transmitters of small output power operating at the same frequency can be used.
- Fig. 2.6 shows the concept of cellular frequency reuse, where cells labelled with the same letter use the same group of RF channels. The frequency reuse plan is overlaid upon a map to indicate where different frequency channels are used.
- The hexagonal shaped cells are shown to model the coverage areas and the base station transmitters are depicted as either being in the centre of the cell (centre excited cells) or on three of the six cell vertices (edge-excited cells). Normally, omnidirectional antennas are used in the centre-excited cells and sectored directional antennas are used in the corner (edge)-excited cells.

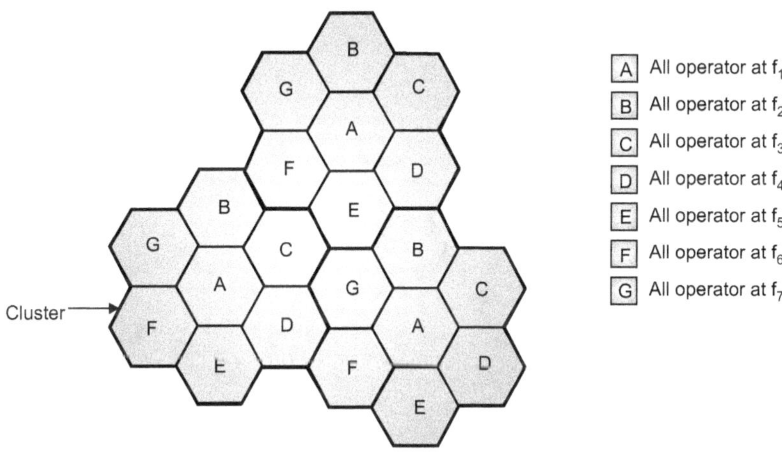


**Fig. 2.6 : Cellular frequency reuse concept**

### 2.6.5 Frequency Reuse Scheme

- Frequency reuse schemes have been proposed to improve the spectral efficiency and signal quality. The different schemes provide different trade offs between resource utilization and quality of service. The classical reuse - 3 scheme proposed for GSM systems offers a protection against intercell interference.
- However, only a third of the spectral resources are used within each cell. In the reuse-1 scheme in which all the resources are used in every cell, interference at the cell edge may be critical. Fractional reuse schemes are a special case, where the reuse-1 is applied to users with good quality (close to the station) and reuse-3 (or higher reuse factor) to users with proper quality (close to the cell edge).

### 2.6.6 Analysis

- Consider a cellular system which has a total number of F full duplex channels available for use in a cluster. If each cell is allocated a group of G channels (G < F) and if the F channels are divided among N cells into unique and disjoint channel groups which each have the same number of channels. The total number of available radio channels can be expressed as,

$$F = GN \qquad \dots (2.1)$$

- *The N cells which collectively use the complete set of available frequencies is called a cluster.*

## 2.6.7 Frequency Reuse Factor

- **The frequency reuse factor is the rate at which the same frequency can be used in the network.**
- *The reciprocal of number of calls 'k' which cannot use the same frequencies for transmission.*

$$\therefore \quad \text{FRF} = \frac{1}{k} \quad \ldots (2.2)$$

- The channel capacity is directly proportional to the number of times, a cluster is duplicated in a given geographical service area. Each cell within a cluster is only assigned 1/N of the total available channels in the cellular system.
- The number of users is equal to frequency reuse factor (FRF).

$$\therefore \quad \text{FRF} = \frac{N}{C} \quad \ldots (2.3)$$

where,
N – Total number of full duplex channels in a given area
C – Total number of full duplex channels in a cell
FRF – Frequency reuse factor

## 2.6.8 Frequency Reuse Distance

- Fig. 2.7 illustrates the frequency reuse patterns for k = 4, 7, 12 and 19. The minimum distance which allows the same frequency to be used will depend on many factors such as the number of co-channel cells in the vicinity of centre cell, the type of geographic terrain contour, the antenna height and the transmitted power at each cell site.

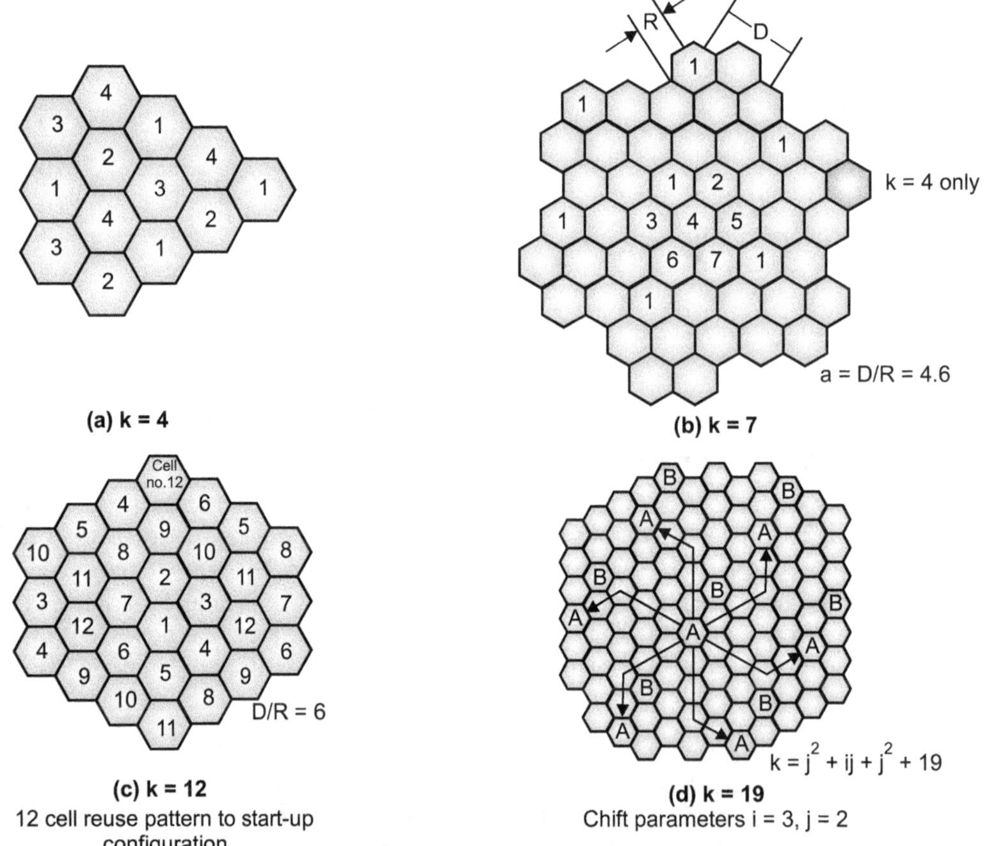


**Fig. 2.7 : Frequency reuse patterns for k = 4, 7, 12 and 19**

- **Cells with the same frequency band must have a minimum distance from each other, is known as frequency reuse distance.**

- The frequency reuse distance 'D' is given by, $D = \sqrt{3\,k} \cdot R$ ... (2.4)

  where, k = Number of cells in frequency reuse pattern

  and R = Radius of geographic coverage area

- If all the cell sites transmit the same power, then 'k' increases and the frequency reuse distance 'd' also increases. This increased distance 'D' reduces the chance that co-channel interference may occur. Theoretically, a large value for 'k' is desired. However, the total number of allocated channels is fixed. When the value of 'k' is too large, the number of channels assigned to each of 'k' cells become small.
- **The minimum distance which allows the same RF channels in the cellular mobile radio system is called frequency reuse distance.**
- The number of frequency response cell 'k' depends on carrier to interference (C/I) ratio. It is expressed as,

$$k = \sqrt{\frac{2}{3} \cdot \frac{C}{I}} \quad \text{... (2.5)}$$

- If the ratio C/I is to be high, then the more frequency reuse cells are required.
- The co-channel interference is a function of a parameter 'Q' and it is expressed as,

$$Q = \frac{D}{R} \quad \text{... (2.6)}$$

- The parameter 'Q' is called the co-channel interference reduction factor.

## 2.6.9 Frequency Reuse Channels

- For simultaneous communication in both directions i.e., full duplex operation and the Radio Frequency (RF) channel uses a pair of frequencies. It uses one frequency for each direction of transmission in the geographic zone of a cellular system.
- Consider Fig. 2.8 to understand the concept of frequency reuse channel of the cellular mobile radio system.
- Let us consider a RF channel corresponding to frequency $F_1$ used in one geographic zone (cell) to call the cell $C_1$ with a coverage radius 'R'.
- This RF channel can be used in another cell $C_2$ with the same coverage radius R at a distance 'D' away.

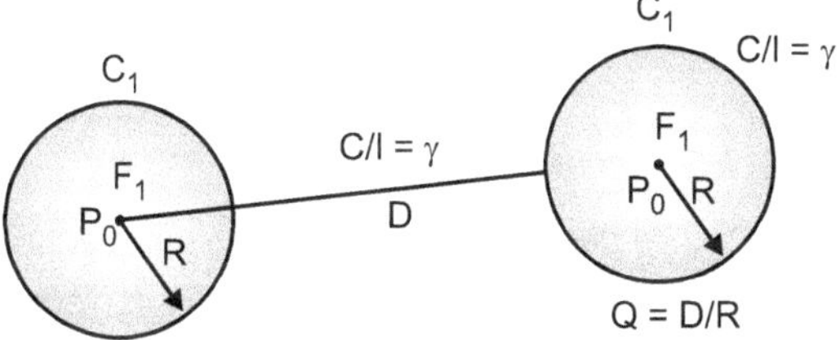


**Fig. 2.8 : Ratio of D/R**

- Thus the RF channel $F_1$ is used by the two cells in different geographic locations. Therefore the frequency reuse allows two different users in different geographic locations to use the same frequency channel simultaneously.

## 2.6.10 Frequency Reuse Patterns

- Depending on the cluster size (N), the frequency reuse patterns can be drawn as under :

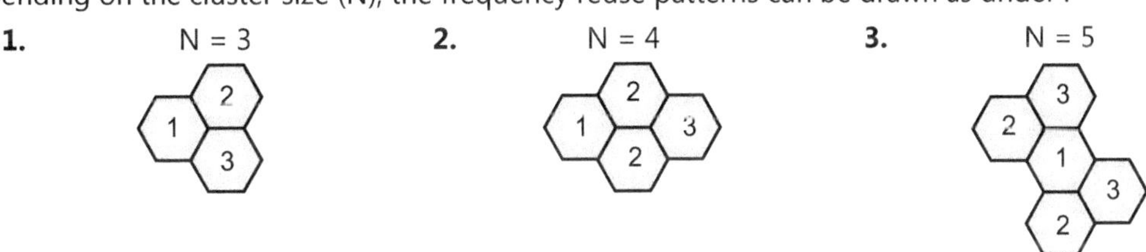


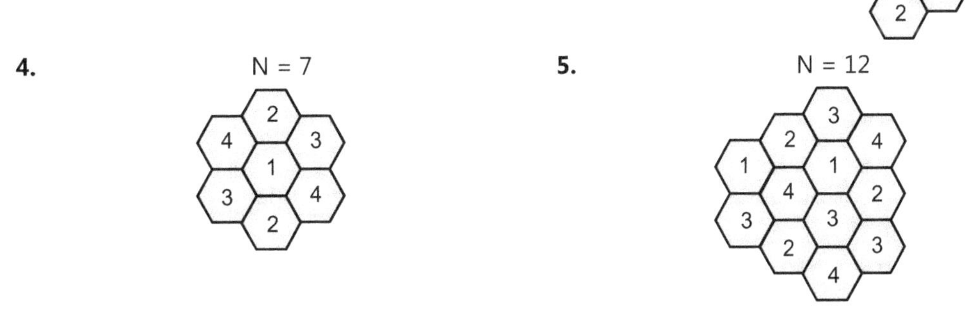


**Fig. 2.9 : Frequency reuse patterns**

### 2.6.11 Advantages

1. It increases the number of conversations of the cellular system with available cellular channels.
2. It increases the spectrum efficiency of cellular systems.
3. It increases the frequency spectrum.
4. It keeps the co-channel interference level within the tolerable limits.
5. It reduces the minimum height of the transmitting antenna, because each antenna has to cover a small geographical area.
6. It allows the use of transmitters of small output power operating at the same frequency.

### 2.6.12 Channels in a Cluster

- Frequency reuse concept can be expressed mathematically by considering a cellular system with a fixed number of full duplexer radio channels available in a given geographic area.
- Each service area is divided into clusters and allocated a group of channels, which is divided among N-cells in a unique and disjoint channel grouping, where all cells have the same number of channel, but do not necessarily cover the same size geographical area.
- The total channels in a cluster is given are,

$$F = GN \quad \text{... (2.7)}$$

where, N = Number of cells in a cluster

G = Number of full duplex channel in a cluster

- When cluster is duplicated 'm' times within a given service area, total number of full-duplex radio channels is given as,

$$C = mF = mGN \quad \text{... (2.8)}$$

where, m = Number of cluster in a given area

C = Total radio channel capacity in a given service area

### 2.6.13 Demerits

1. The design of the cellular system becomes very crucial and hence it is to be properly designed.
2. It may give rise to co-channel interference, if the cellular system is not properly designed.

- Thus, the co-channel interference is the major limitation of the frequency reuse cellular system.

### 2.6.14 Important Terms

Some of the important terms related to the concept of frequency reuse are defined as under :

**1. Frequency reuse :**

- The design process of selecting and allocating RF channel groups for all of the cellular base stations within a cellular system is called frequency reuse.
- It refers to the use of radio channels operating on the same frequency, to cover different geographical areas, that are physically separate from each other.

**2. Cluster :**

- The number of cells that collectively use all the available frequencies is called a cluster. It is denoted by a letter N.

**3. Cluster size :**

- The number of cells in a cellular system is called a cluster size (N).
- Its value is typically equal to 4, 7, 12 or 19. It should be as small as possible.

**4. Capacity :**

- The total number of duplex channels in a cellular system is called capacity of a cellular system. It is denoted by a letter C.

**5. Frequency reuse factor (ratio) :**

- The reciprocal of the cluster size (N) in a cellular system is called frequency reuse factor.
- Frequency reuse factor = 1. It should be as large as possible.

6. **Frequency reuse distance :**
   - The minimum distance which allows the same RF channels in a cellular system is called frequency reuse distance. It is denoted by a letter D.

## 2.7 CELLULAR SYSTEM INTERFERENCE

### 2.7.1 Introduction

- In modern times, communication technologies serve as drivers of social, economical and political developments. But interference in communication networks comes as undesirable nuisance to the system's quality of service.
- Despite of efforts to reduce this interference, challenges of interference still abound in communication networks. Interference poses a major problem in GSM networks for service providers as it reduces the quality of service.

### 2.7.2 Concept

- The cellular system interference occurs when two waves that left one source and travelled by different paths arrive at a point. This happens very often in high frequency (sky wave or microwave, space wave) propagation. In cellular mobile radio systems, the interference takes place.
- **The cellular system interference is the major limiting factor in the performance of cellular mobile radio systems.**
- It is more severe in urban areas due to the greater RF noise factor and also the large number of mobile stations and mobile phone units. It has been recognized as a major bottleneck in increasing capacity and is often responsible for dropped cells.

### 2.7.3 Sources

- The various sources of interference in the cellular radio systems are as under :
  1. Another mobile phone unit in the same cell.
  2. Call in progress in a neighbouring cell.
  3. Other base stations operating in the same frequency band.
  4. Any other non cellular radio system which inadvertently leaks energy into the cellular frequency band.

### 2.7.4 Effects

- The cellular system interference involved in the cellular mobile radio systems causes the following effects on the various signals :
  1. The cellular system interference on the voice channels causes cross-talk when the subscriber hears the interference in the background due to an undesired transmission.
  2. The cellular system interference on control channels leads to missed and blocked calls due to errors in the digital signaling.

### 2.7.5 Types

- The following types of interference may be associated with cellular networks :
  1. Self interference
  2. Multiple access interference
  3. Co-channel interference
  4. Adjacent channel interference

### 2.7.6 Methods for Reducing Interference

- The methods for reducing the interference (co-channels or adjacent channels) in the cellular mobile radio systems are as under :

1. **A good Frequency Management Chart :**
   - In a frequency management chart, there are 21 sets of channels in the chart.
   - In each channel set, the neighbouring frequency is 21 channels away.
   - No interference can be caused within a set of 16 channels.

2. **An Intelligent Frequency Assignment :**
- In order to assign 21 sets in a k = 7 frequency reuse pattern and avoid the interference problems from adjacent channel or co-channel interference, intelligent frequency assignment in real time is needed.

3. **A Proper Frequency Among a Set Assigned to a Particular Mobile Unit :**
- Depending on current situation, some *idle channels* may be noisy, some may be quite and some may be vulnerable to channel interference.
- These factors should be considered in assignment of frequency channels.

4. **Design of an Antenna Pattern on the Basis of Direction :**
- In some directions, a strong signal may be needed; in other directions no signal may be needed.
- The design tool should include the findings of signal requirements on the basis of antenna directions.

5. **Tilting-antenna Patterns :**
- To confine the energy within a small area, we may use an umbrella - pattern omnidirectional antenna or downward tilting directional antenna.

6. **Reducing the Antenna Height :**
- We can use this method because reducing interference is more important than radio coverage.

7. **Reducing the Transmitted Power :**
- In certain circumstances reducing transmitted power can be more effective in eliminating interference than reducing the height of the antenna.

8. **Choosing the Cell-site Location :**
- The propagating prediction model can be used to select cell-site locations for eliminating interference.

## 2.8 CO-CHANNEL INTERFERENCE

### 2.8.1 Introduction

- The name co-channel interference is a bit deceiving, since it is not interference in the traditional sense. It is actually the 802.11 protocol doing what it does best, serving clients to the best of its ability and surviving. Co-channel interference occurs between two access points that are on the same frequency channel.

### 2.8.2 Definition

- **Co-channel interference is a cross-talk from two different ratio transmitters using the same channel.**
- *The channel having the same set of frequencies in a given coverage area is called a co-channel.*
- *The interference between the signals of co-channel cells that repeatedly use the same set of frequencies in the cellular mobile radio system is called co-channel interference.*

### 2.8.3 Cause

- Co-channel interference can be caused by many factors from weather conditions to administrative and design issues. Co-channel interference may be controlled by various radio resource management schemes. The reason that you should care is that co-channel interference can severely affect the performance of wireless LAN (WLAN). The spectrum that is available for the deployment of Wi-Fi is limited.

### 2.8.4 Description

- The co-channel interference occurs in the cellular phone systems, when the same frequency channel is used repeatedly in co-channel cells. We cannot reduce the co-channel interference by simply increasing the transmitter output power.
- In fact, increasing the transmitter output power will increase the co-channel interference in the cellular phone system. The co-channel interference in the cellular phone system can be reduced by separating the co-channel cells physically by a minimum distance to provide sufficient isolation due to propagation.
- If all the cells are of the same size and all the base stations are transmitting equal amount of power, the co-channel interference ratio is independent of the transmitted (output) power, but it becomes a function of the radius of a cell (R) and the distance D between centres of the nearest co-channel cells. If we increase the ratio (D/R), then the separation between the co-channel cells relative to the coverage distance of a cell increases

thereby it reduces the co-channel interference in the cellular phone system. The parameter Q called as the co-channel reuse ratio is related to the D/R ratio and also the cluster size N for the hexagonal geometry as under :

$$Q = \frac{D}{R} = \sqrt{3N} \qquad \text{... (2.9)}$$

- A small value of Q provides the larger system capacity since the cluster size N is small, whereas a large value of Q provides the smaller system capacity since the cluster size N is large. Thus the large value of Q reduces the level of co-channel interference and improves the transmission quality of the cellular phone system.
- Hence, the selection of co-channel reuse ratio Q is based on the following two factors :
  1. System capacity
  2. Co-channel interference
- The co-channel interference determines the link performance, which in turn dictates the frequency reuse plan and the overall capacity of the cellular phone systems.

### 2.8.5 Methods for Reducing Co-channel Interference

- The reduction in co-channel interference in a cellular mobile system is always a challenging problem. The important methods for reducing the co-channel interference are as under :
  1. Increasing the separation between two co-channel cells.
  2. Using directional antennas at the base station.
  3. Lowering the antenna heights at the base station.
- The use of directional antenna in each cell can serve the following two purposes :
  1. Further reduction of co-channel interference, if the interference cannot be eliminated by a fixed separation of co-channel cells.
  2. Increasing the channel capacity when the traffic increases.

### 2.8.6 Effects

1. The number of frequency reuse increases the system efficiency and hence decreases the number of channels per cell.
2. It weakens the reception level at the mobile phone handset.
3. It effects voice quality.
4. It effects the overall cellular system capacity and thereby it reduces the system capacity.

## 2.9 ADJACENT CHANNEL INTERFERENCE

### 2.9.1 Definition

- **The interference resulting from signals which are adjacent in frequency to the desired signal is called adjacent channel interference.**
- It results from imperfect receiver filters which allow nearby frequencies to leak into the pass band. The receiver attempts to receive a base station on the desired channel.

### 2.9.2 Basic Principle

- The basic principle of the adjacent channel interference is illustrated in Fig. 2.10.
- The imperfect filters of a receiver allow the frequencies from the adjacent channels to leak in the pass band of the desired signal.
- This phenomenon results the adjacent channel interference in the cellular phone system.

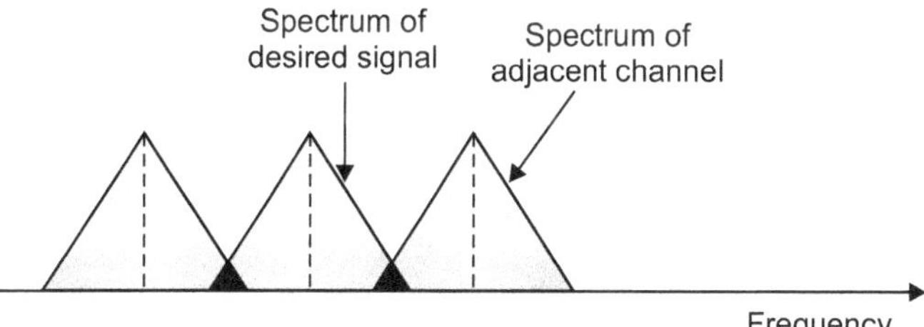


**Fig. 2.10 : Adjacent channel interference**

- The adjacent channel interference will be serious, if the adjacent channel user is transmitting in very close range to the subscriber's receiver. The receiver always attempts to receive a base station on the desired channel.

### 2.9.3 Description

- The adjacent channel interference is a broad term. It includes the following interferences :
  1. Next-channel (the channel next to the operating channel) interference.
  2. Neighbouring channel (more than one channel away from the operating channel) interference.
  3. Next end - far end interference.

  These interferences are described as under :

**1. Next Channel Interference :**

- The next-channel interference affecting a particular mobile phone unit cannot be caused by transmitters in the common cell site, but must originate at several other cell sites. This is because any channel combined at the cell site must combine the selected channels, normally 21 channels (630 kHz) away, or at least 8 to 10 channels away from the desired one.
- Therefore next channel interference will arrive at the mobile phone unit from other cell sites, if the system is not designed properly. Also a mobile phone unit initiating a call on a control channel in a cell may cause interference with the next control channel at another cell site. The methods for reducing this next-channel interference use the receiving end.
- The channel filter characteristics are a 60 dB/octave slope in the voice band and a 24 dB/octave fall off outside the voice band region. If the next-channel signal is stronger than 24 dB, it will interface with the desired signal. The filter with a sharp falloff slope can help to reduce all the adjacent channel interference, including the next-channel interference.

**2. Neighbouring Channel Interference :**

- The channels which are several channels away from the next channel will cause interference with the desired signal. Usually a fixed set of serving channels is assigned to each cell sites. If all the channels are simultaneously transmitted at one cell site antenna, a sufficient amount of band isolation between channels is required for a multi-channel combiner to reduce intermodulation products.
- This requirement is not different from other non-mobile radio systems. Assume that band separation requirements can be resolved, for example, by using multiple antennas instead of one antenna at the cell site. What channel separation would be needed to avoid adjacent channel interference ?
- Another type of adjacent channel interference is unique to the mobile radio system. In the mobile radio system, most mobile units are in motion simultaneously. Their relative positions change from time to time. In principle, the optimum channel assignments that avoid adjacent channel interference must also change from time to time.

**3. Near end - Far end Interference :**

- If motor vehicles in a given cell are moving, then some mobile phone units are close to the cell site and some are not. The close in mobile phone unit has a strong signal which causes adjacent-channel interference as shown in Fig. 2.11 (a). In this situation, near end - far end interference can occur only at the reception point in the cell site.
- If a separation of five channel bandwidth (5B) is needed for two adjacent channels in a cell in order to avoid the near end-far end interference. It is then implied that a minimum separation of 5B is required between each adjacent channel used with one cell. The adjacent channel interference can also occur between two systems in a duopoly-market system.

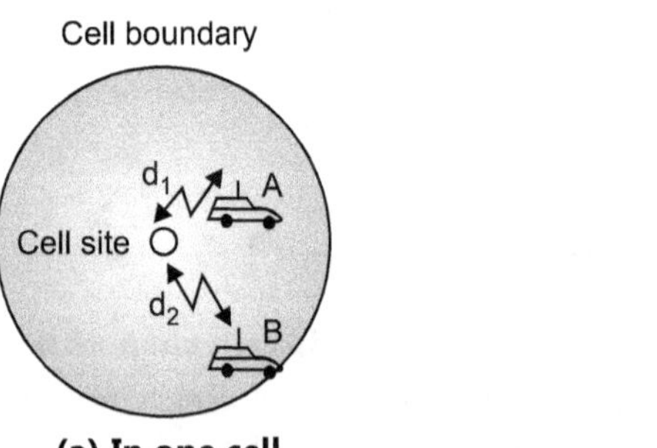


**(a) In one cell**

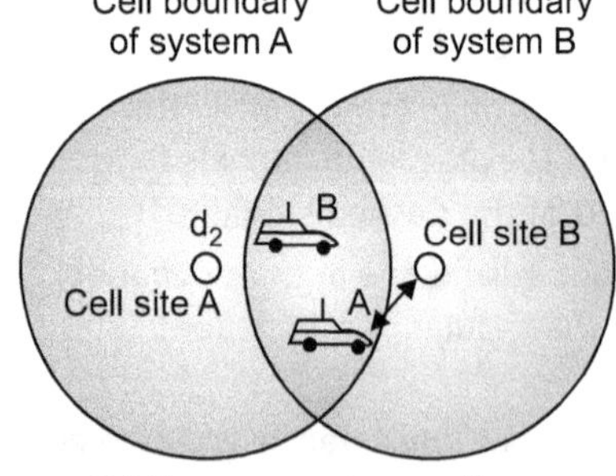


**(b) In two-system cells**

**Fig. 2.11 : Near end-far end (ratio) interference**

- In this situation, adjacent channel interference can occur at both the cell site and the mobile phone unit.
- For instance, mobile phone unit A can be located at the boundary of its own home cell A in system A but very close to cell B of system B as shown in Fig. 2.11 (b). The other situation would occur, if mobile phone, unit B was at the boundary of cell B of system B, but very close to cell A of system A. In Fig. 2.11 (b), the solid arrow indicates that interference may occur at cell site A and the dotted arrow indicates the interference may occur at mobile phone unit A.
- The same interference will be introduced at cell site B and mobile unit B. The two causes of near end-far end interferences are as under :
  1. Interference caused on the set up channels.
  2. Interference caused on the voice channels.
- The adjacent channel interference can be minimised through careful filtering and channel assignments. Since, each cell is given only a fraction of the available channels, a cell need not be assigned channels which are all adjacent in frequency.
- By keeping the frequency separation between each channel in a given cell as large as possible, the adjacent channel interference may be reduced considerably. Some channel of allocation schemes also prevent a secondary source of adjacent channel interference by avoiding the use of adjacent channels in neighbouring cell sites. In practice, the base station receivers are preceded by a high Q cavity filter in order to reject adjacent channel interference. The adjacent channel interference can be reduced by the frequency assignment.

### 2.9.4 Reduction Techniques

- The adjacent channel interference in a cellular phone system can be reduced by taking proper precautions as under :
  1. Careful filtering
  2. Careful channel assignment
- The adjacent channels should not be overlapped. But there should be adequate frequency separation between the adjacent channels in a cell.

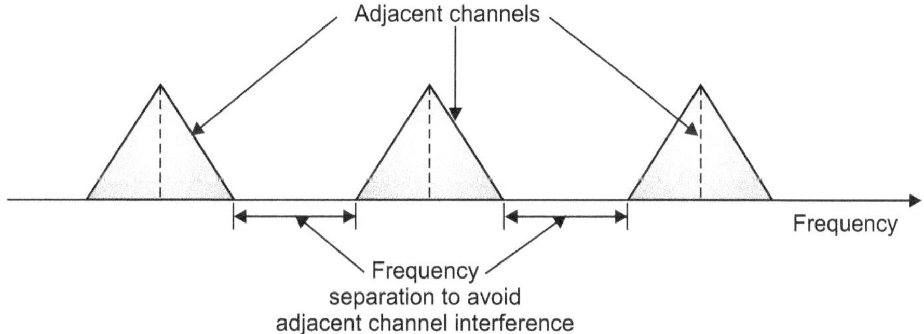


**Fig. 2.12 : Frequency separation to avoid adjacent channel interference**

- If the frequency reuse factor is large (or if the cluster size N is small), then the separation between adjacent at the base station may not be sufficient. This may increase the adjacent channel interference. In practice, the base station receivers use high Q cavity filters to minimize the adjacent channel interference.

## 2.10 REDUCTION FACTOR

### 2.10.1 Introduction

- The importance of estimating reduction factor R originates in the need for the direct driving elastic spectra. Brief steps for deriving an elastic spectrum are : Use approximate R factor model to estimate R factor, and then use the estimated R-factor to scale the elastic spectra. This is an approximate approach, but as expected, the error produced by the approach is tolerable in engineering practice.
- In recent years, near fault forward-directivity (FD) ground motion has attracted attention because it has caused an amount of structural damage in recent strong earthquake events. To obtain the inelastic spectra

of forward-directivity ground motions, we should know, which existing R-factor model to appropriate for use, as most existing R-factor models were developed mainly from far-field records.

- In this study, we have two objectives as under :
  1. To verify which one of existing R-factor models is suitable for use in estimating R-factor for FD ground motions.
  2. To develop a new R-factor model by using FD records and accounting for the effect of site conditions.
- For these purposes, two groups of FD records, recorded on soil and rock sites, are selected carefully from world-wide strong motion databases, six ductility factors from 2 to 8 are used as target ductility factors, the longest period used is 4s and three hysteritic models, perfectly elastic-plastic. Bilinear stiffness degrading models are used to represent structural load-deformation mechanism.
- Reduction factor of the strength is found by dividing the value of elastic strength by the yield strength. It is represented using the R-factor. Usually, the significance of evaluating R-factor emerges in the requirement for directly obtaining the inelastic range.

### 2.10.2 Definition

- **Reduction factor of the strength is defined as the ratio of elastic strength to yield strength.** It is denoted by a letter R.

## 2.11 OMNI-DIRECTIONAL ANTENNA SYSTEM

### 2.11.1 Antenna Concept

- An antenna gives the wireless system three fundamental properties such as *gain, direction* and *polarization.* Gain is a measure of increase in power. Gain is the amount of increase in energy that an antenna adds to a radio frequency (RF) signal. Direction is the shape of the transmission pattern.
- As the gain of a directional antenna increases, the angle of radiation usually decreases. This provides a greater coverage distance, but with a reduced coverage angle. The coverage area or radiation or radiation pattern is measured in degrees. These angles are measured in degrees and are called **bandwidths.**
- An antenna is a passive device, which does not offer any added power to the signal. Instead, an antenna simply redirects the energy, it receives from the transmitter. The redirections of this energy has the effect of providing more energy in one direction, and less energy in all other directions.
- Beamwidths are defined in both horizontal and vertical plains. Beamwidth is the angular separation between the half power (3 dB) points in the radiation pattern of the antenna in any plane. Therefore, for an antenna you have horizontal beamwidth and vertical beamwidth.
- Antennas are related to comparison to isotropic or dipole antennas. An isotropic antenna is a theoretical antenna with a uniform three dimensional radiation pattern (similar to a light bulb with no reflection).
- In other words, a theoretical isotropic antenna has a perfect 360° vertical and horizontal beamwidth or a spherical radiation pattern. It is an ideal antenna, which radiates in all directions and has a gain of 1 (0 dB), i.e. zero gain and zero loss. It is used to compare the power level of a given antenna to the theoretical isotropic antenna.

### 2.11.2 Definition

- **In radio communication, an omni-directional antenna is a class of antenna, which radiates a equal radio power in all directions perpendicular to an axis (azimuthal directions), with power varying with angle to the axis (elevating angle), declining to zero on the axis.**
- *An omni-directional antenna is a wireless transmitting or receiving antenna that radiates or intercepts radio frequency (RF) electromagnetic fields equally well in all horizontal directions in a flat, two dimensional geometry plane.*

### 2.11.3 Classification

- Antennas can be broadly classified into two categories as under :
  1. Omnidirectional antenna.
  2. Directional antenna.

### 2.11.4 Directional Antenna

- Unlike isotropic antennas, dipole antennas are real antennas. The dipole radiation pattern is 360° in the horizontal plane and approximately 75° in the vertical plane (this assumes the dipole antenna is standing vertically) and resembles a doughnut in shape. Because the beam is slightly concentrated, dipole antennas have a gain over isotropic antennas of 2.14 dB in the horizontal plane.
- Dipole antennas are said to have a gain of 2.14 dB; which is in comparison to an isotropic antenna. The higher the gain of the antennas, the smaller the vertical beamwidth is.

### 2.11.5 Omni-directional Antenna

- Omni-directional antennas have a similar radiation pattern. These antennas provide a 360° horizontal radiation pattern. These are used when coverage is required in all directions (horizontally) from the antenna with varying degrees of vertical coverage.
- Polarization is the physical orientation of the domain on the antenna that actually exists the RF energy. An omni-directional antenna, for example, is usually a vertical polarized antenna. Directional antenna focus the RF energy in a particular direction. As the gain of directional antenna increases, the coverage distance increases, but the effective coverage angle decreases.
- For directional antennas, the lobes are pushed in a certain direction and little energy is there on the backside of the antenna. The higher the gain of the antenna, the higher the front-to-back ratio is. A good antenna front-to-back ratio is normally 20 dB.
- Common types of low-gain omni-directional antennas are whip-antenna, rubber-ducky antenna, ground plane antenna, vertically oriented dipole antenna, discona antenna, mast radiator, horizontal loop antenna, (sometimes known colloquibally as a *circular aerial* because of the shape) and the hole antenna.
- Higher gain omni-directional antennas can also be built. *'Higher gain'* in this case means that the antenna radiates less energy at higher and lower elevation angles and more in the horizontal direction. High gain omni-directional antennas are generally realized using *collinear dipole arrays*. These consist of multiple half wave dipoles mounted colinearly (in line), fed in phase.
- The Coaxial Collinear (COCO) antenna uses transposed coaxial sections to produce in phase half-wavelength radiators. A *Franklin Array* uses short U-shaped half wavelength section, whose radiation cancels in the far-field to bring each half-wavelength dipole section into equal phase. Another type is the Omni-directional Microstrip Antenna (OMA).

### 2.11.6 Desired C/I from a Normal Case

- We can obtain in decibels (dB), the Carrier to Interference ratio (C/I) by subtracting the result obtained from $f_2$ from the result obtained from $f_1$ (carrier minus interference C–I), and the Carrier-to-Noise ratio (C/N) by subtracting the result obtained from $f_1$ from the result obtained from $f_2$ (Carrier minus Noise C–N). Four conditions should be used to compare the results.
  1. If the Carrier-to-Interference ratio (C/I) is greater than 18 dB throughout most of the call, the system is properly designed.
  2. If C/I is less than 18 dB and C/I is greater than 18 dB in some areas, there is a cochannel interference.
  3. If both C/N and C/I are less than 18 dB and C/N – C/I in a given area, then there is a coverage problem.
  4. If both C/N and C/I are less than 18 dB and C/N > C/I in a given area, then there is a coverage problem and cochannel interference.

## 2.12 CELL SPLITTING

### 2.12.1 Need

- The motivation behind implementing a cellular mobile (radio) system is to improve the utilization of spectrum efficiency. When the call traffic density in an area starts to build up and the frequency channel $F_i$ in each cell $C_i$ cannot provide enough mobile cells to increase the capacity of the cellular systems.
- Therefore, there is a necessity of cell splitting to improve the capacity of a cellular system by increasing the number of reused channels.

### 2.12.2 Concept

- The cell splitting achieves the capacity improvement by essentially rescaling the cellular system. By decreasing the cell radius R and keeping the co-channel reuse ratio D/R unchanged, cell splitting increases the number of channels per unit area.
- **The cell splitting is the process of subdividing a congested cell into smaller cells with its own base station, having the corresponding reduction in the antenna heights and the transmitted power.**
- Thus, cell splitting allows the orderly growth of the cellular system. It increases the number of base stations in order to increase the capacity of a cellular system. In this technique, the original cells are sub-divided to form the new cells of similar size called as microcells.
- The microcells have a smaller radius than the original cells. The microcells are installed between the existing cells. This increases the system capacity of cell traffic many fold due to the additional number of channels per unit area.

### 2.12.3 Description

- Imagine, if every cell as shown in Fig. 2.13 was reduced in such a way that the radius of every cell was cut in half. The base stations are placed at corners of the cells and the area served by base stations A is assumed to be saturated with traffic (i.e., the blocking of the base station A exceeding the acceptable rates).
- The new base stations are therefore needed in the region to increase the number of channels in the area to reduce the area served by the single base station.
- Note that the original base station A is surrounded by six new microcells. In the example shown in Fig. 2.13, the smaller cells are added in such a way as to preserve the frequency reuse plan of the system. For example, the microcell base station labelled G placed half wax between two larger stations utilizing the same channel set G. This is also the case for other microcells shown in Fig. 2.13.

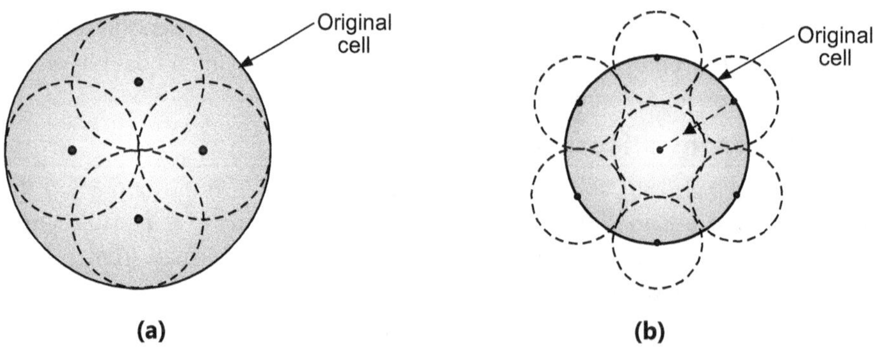


**Fig. 2.13 : Concept of cell splitting**

$$\text{New cell radius} = \frac{\text{Old cell radius}}{2}. \quad \text{So, new cell area} = \frac{\text{Old cell area}}{4}$$

- Let each new cell carry the same maximum traffic load of the old cell, then in theory,

$$\frac{\text{New traffic load}}{\text{Unit area}} = 4 \times \frac{\text{Traffic load}}{\text{Unit area}}$$

- In order to cover the entire service (coverage) area with smaller cells, approximately four times as many cells would be required. This can be easily shown by considering a circle with radius R. The area covered by such a circle is four times as large area covered by a circle with radius R/2. This increased number of cells would increase the number of clusters over the coverage region, which in turn would increase the number of channels and thus capacity in the coverage area.
- In practice, not all cells are split at the same time. It is often difficult for service providers to find real estate that is perfectly situated for cell splitting. Therefore different cell size will exist simultaneously. In such situation, special care needs to be taken to keep the distance between the co-channel cells at the required minimum and hence the channel assignments become more complicated.
- As the cell splitting continues, the general formula can be expressed as,

$$\text{New traffic load} = 4^n \times \text{Traffic load of start-up cell}$$

- For n = 4, the original start-up cell will split 4 times. The new traffic load will be 256 times larger than the traffic load of the start-up cell.

## 2.12.4 Techniques

- There are two types of cell splitting techniques, namely :
  1. Permanent cell splitting
  2. Real time (i.e. dynamic) cell splitting.

## 2.12.5 Permanent Cell Splitting

- The selection of small cell sites is a tough job.
- Therefore the installation of every new split cell has to be planned ahead of times, the number of channels, the transmitted power, the assigned frequencies, the choosing of the cell site selection and the traffic load consideration should all be considered.
- When the splitting is ready, the actual service cut-over should be set at the lowest traffic point, usually at midnight on a weekend.
- It will drop a few cells because of this cut-over, assuming that the downtime of the cellular radio system is within two hours. The antenna can be mounted on a monopole or erected by a mastless arrangement.
- However, these splitting can be easy to handle as long as the cut-over from large cells to small cells takes place during a low traffic period.

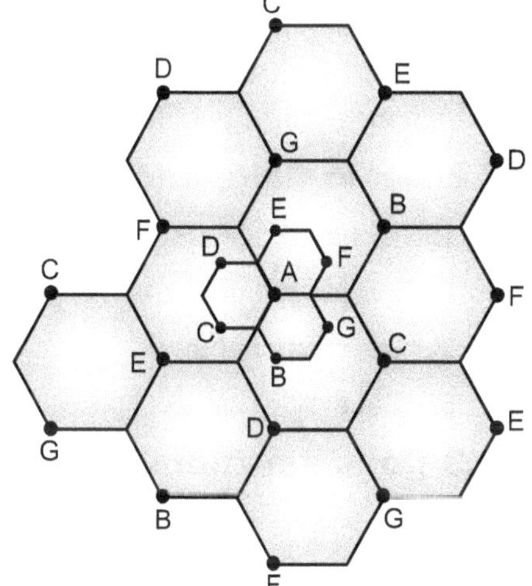

**Fig. 2.14 : Permanent splitting**

- The frequency alignment should follow the rule based on the frequency reuse distance ratio Q with the power adjusted.

## 2.12.6 Dynamic Cell Splitting

- This technique is based on utilizing the allocated spectrum efficiency in real time. The algorithm for dynamically splitting cell sites is a tedious job since we cannot afford to have one single cell unused during cell splitting at heavy traffic hours.
- In many situations, such as traffic jams at football stadiums after a game, in traffic jams resulting from automobile accidents and so on, the idle small cell sites (inactive ones) may be rendered operative in order to increase the cell's traffic capacity. The cell splitting should be proceeded gradually over a cellular operating system to prevent dropped cells.

## 2.12.7 Limitations

- The size of a splitting cell is dependent on the two following factors :
  1. **The radio aspect :** The size of a small cell is dependent on how well the coverage pattern can be controlled and how accurately vehicle locations would be known.

2. **The capacity of the switching processor :** The smaller the cells, the more hand-offs will occur and the more cell splitting process is needed. This factor, of a switching processor, is a larger factor than the handling of coverage areas of small cells.

### 2.12.8 Effects

- When the cell splitting is caused to maintain the frequently reuse distance ratio Q in a cellular system, there are two considerations. These considerations are as under :
  1. The cell splitting affects the neighbouring cells. The splitting cells causes an unbalance situation in power and frequency reuse distance and makes it necessary to split small cells in neighbouring cells. This phenomenon is the same as a ripple effect.
  2. The certain channels are used as barriers. To the same extent, large and small cells can be isolated by selecting a group of frequencies which will be used only in the cells located between the large cells on one side and the small cells on the other side, in order to eliminate the interference being transmitted from the large cells to the small cells.

## 2.13 CONSIDERATIONS OF COMPONENTS

### 2.13.1 Introduction

- There are various cellular systems in the world, such as the GSM and CDMA. The design of these cellular systems are complicated, but the architecture of most cellular systems can be broken down into six basic components.
- It aims to provide an introductory guide to the architecture of a typical cellular system and identify the six basic components found in most cellular systems. It offers an insight towards how a cellular system is designed, although different cellular systems may have variations in their own implementations.

### 2.13.2 Basic Components

- The six basic components that can be found in most cellular systems are as under :
  1. Mobile Station (MS)
  2. Air interference standard
  3. Base Station (BS)
  4. Database
  5. Gateway
  6. Security mechanism.

### 2.13.3 Considerations

- Many design related problems in engineering are ambiguous and difficult to handle, the engineers face the task of finding the solution to a problem that involves the considerations of multiple contradicting objectives and constraints. A solution that provides an appropriate balance between all the objectives and meets the given constraints is found only after an exhaustive trial-and process.
- However, the optimality of the solution is not known and the quality of the solution is usually not guaranteed. On the other hand, in the increasing competitive wireless industry, there is a need to provide good quality solution consistently. An optimization approach offers the means to achieve such goal. In order to use optimization to solve a problem, we need to consider the following steps :
  1. To identify the variables and describe the problem quantitatively.
  2. After the problem is clearly defined, the next step is to construct a mathematical model.
     - The mathematical model of a problem is a system of equations that relates the essence of the problem to the decision variables.
     - If there are 'n' related quantifiable decisions to be made, then they are defined as no decision variables.
     - The dimensions of a problem are the number of decision variables that the problem considers, while the solutions to the problem.

3. In wireless systems engineering, there are numerous types of objective that can be associated with an optimization problem.
   - Some of these objectives are essentials, such that they must be satisfied, while the others are only desirable. An example, would be the minimization of the number of base stations required in a wireless system.
   - These type of objectives are usually combined and expressed as a mathematical function of the decision variables called the objective function.
4. This is the last step of the network planning procedure. It constraints during the network trial period and continues after opening the commercial service and during the network expansion. The aim of this process is to evaluate and maximize the quality of service in the network with the corresponding set of quality criteria.

## 2.13.4 Deciding Factors

- The factors that need to be taken into consideration in the planning phase include the following :
  1. Area of coverage needed.
  2. Sites required for the area.
  3. Number of sites needed based on capacity and coverage requirements.
  4. Dimension of sites.
  5. Types of clutters (urban, suburban, open water, etc.)
  6. Identification of search areas covering all clutter types.
  7. Survey sites with reference to clutter heights, vegetation levels, obstructions etc.
  8. Building strengths and other civil requirements.
  9. Propagation tests.
  10. Coverage probabilities.
  11. Find coverage map.

## *Practice Questions*

1. Draw the schematic diagram of a basic cellular system and describe it in brief.
2. What do you mean by frequency reuse in cellular system ?
3. Describe the concept of frequency reuse.
4. Draw two frequency patterns.
5. Draw frequency reuse patterns with cluster size 7 and 12.
6. Explain concept of a cell.
7. What is a cell cluster ?
8. What should be geometric shape of a cell ?
9. Why do you need a cellular radio telephone system ?
10. What is a cellular system ?
11. Draw architecture of a cellular system and explain its operating principle.
12. Draw schematic diagram of a basic cellular telephone system and describe it.
13. What is frequency reuse factor ?
14. What is channel cluster ?

15. What is cellular system interference ?
16. Explain basic concept of cellular system interference.
17. What is co-channel interference ?
18. Describe the effect of co-channel interference in communication.
19. How co-channel interference in a cellular system is reduced ?
20. What is adjacent interference in a cellular system ?
21. Describe adjacent interference in a cellular system.
22. Why do you need cell splitting ?
23. Describe adjacent interference in a cellular system.
24. Explain cell splitting.
25. Explain concept of cell splitting using suitable diagram.
26. How cell splitting helps to increase the system capacity ?
27. Explain dynamic cell splitting and permanent cell splitting.
28. What is omni-directional antenna ?
29. What is desired C/I from a normal case in an omni-directional antenna ?

✍✍✍

Chapter 3

# DIGITAL COMMUNICATION THROUGH FADING MULTIPATH CHANNELS

**Syllabus**

Fading channel and its characteristics - Channel modeling, Digital signaling over a frequency, Non-selective slowly fading channel. Concept of diversity branches and signal paths. Combining methods : Selective diversity combining, Switched combining, Maximal ratio combining, Equal gain combining.

## 3.1 DIGITAL COMMUNICATIONS

### 3.1.1 Brief History

- Data (mainly but not exclusively informational) has been sent via non-electronic e.g. optical, acoustic, mechanical means since the advent of communication. *Analog signal* data has been sent electronically, since the advent of *telephone*. However, the first data electromagnetic transmission applications in modern time were *telegraphy* in 1809 and *tele typewriter* in 1906, which are both *digital signals.*
- The fundamental theoretical work in data transmission and information theory by *Harry Nyquist, Ralph Hartlay, Elude Shannon* and others during the early 20th century, was done with these applications in mind. Data transmission is utilized in *computers* in *computer buses* and for communication with *peripheral equipment* via *parallel ports* and *serial ports* such as RS-232 (1969), firmware (1995) and USB (1996).
- The principles of data transmission are also utilized in storage media for *error detection* and *correction* since 1951. Data transmission is utilized in *computer networking* equipments, such as modems (1940), *Local Area Network* (LAN) adapters (1964), repeaters, repeater hubs, microwave links, wireless network access points (1997), etc.
- In telephone networks, digital communication is utilized for transferring many phone calls over the same copper cables or fiber cable by means of *Pulse Code Modulation* (PCM), i.e. sampling and digitization, in combination with *Time Division Multiplexing* (PCM) 1962.
- Telephone exchanges have become digital and software controlled, facilitating many value added services. For example, the first *AXE telephone exchange* was presented in 1976. Since, the late 1980s, digital communication to the end user has been possible using *Integrated Services Digital Network* (ISDN) services. Since, the end of the 1990s, broadbands, access techniques, such as ADSL, cable modems, Fiber-To-The-Building (FTTB) and Fiber-To-The-Home (FTTH) have become widespread to small offices and homes.
- The current tendency is to replace traditional telecommunication services by *Packet Mode communication* such as IP telephony and IPTV. Transmitting analog signals digitally allows for greater *Signal Processing* (PMSP). The ability to process a communication signal means that errors caused by the random processes can be detected and corrected. Digital signals also be *sampled* instead of continuously monitored.
- The *multiplexing* of multiple digital signals is much simpler to the multiplexing of analog signals. Because of all these advantages and recent advances in *whole communication channels* and *solid-state electronics* have allowed scientists to fully realize these advantages, digital communications have grown quickly. Digital communications are edging out analog communication, because of the vast demand to transmit computer data and the ability of digital communications to do so.

- The digital revolution has also resulted in many digital telecommunication applications, where the principles of data transmission are applied. Examples are *Second Generation* (2G) (1991) and later cellular telephony, video conferencing, digital TV (1986), digital radio (1990), telemetry, etc.

### 3.1.2 Introduction

- In the design of large and complex digital systems, it is often necessary to have one device communicate digital information to and from other devices. One advantage of digital information is that it tends to be far more resistant to be transmitted and interpreted errors than information symbolized in an analog medium.
- This accounts for the clarity of digitally encoded telephone connections, compact audio disks and for much of the enthusiasm in the engineering community of digital communications technology. However, digital communications has its own unique pitfalls, and there are multitudes of different and incompatible ways in which it can be sent.
- The analog system would be simple and robust. For many applications, it would suffice for our needs perfectly. But, it is not the only way to get the job done. For the purpose of exposing digital techniques, we will explore other methods of monitoring any process, even though the analog method might be the most practical.
- The analog system as simple as it may be does have its limitations. One of them is the problem of analog is signal *interference* the inductive and capacitive coupling may create a false *noise* signal to be introduced in the D.C. circuit. If this significantly more then it causes the transmission problem.

### 3.1.3 Definition

- *Data transmission* is also known as *data communication* or *digital communication*.
- *Digital communication is the transfer of data (a digital bitstream or a digitized analog signal) over a point-to-point or point-to-multipoint communication channel.*
- Examples of such channel are *copper wires, optical fibers, wireless* communication channels, *storage media* and *computer buses.* The data are represented as an *electromagnetic signal,* such as an *electrical voltage, radiowave, microwave,* or *infrared signal.*
- **Digital communication is the process of devices communicating information digitally.**

### 3.1.4 Need of Digitization

- The conventional methods of communication used analog signals for long distance communications, which suffer from many losses, such as distortion, interference, and other losses including security breach.
- In order to overcome these problems, the signals are digitized using different techniques. The digitized signals allow the communication to become clear and accurate without losses. Therefore, there is a need of digitization in digital communication.

### 3.1.5 Block Diagram

- The block diagram of a basic digital communication system is shown in Fig. 3.1. It consists of the following basic elements :

  1. Source
  2. Source encoder
  3. Channel encoder
  4. Digital modulator
  5. Channel
  6. Digital demodulator
  7. Channel decoder
  8. Input transducer
  9. Output transducer
  10. ADC and DAC
  11. Output signal.

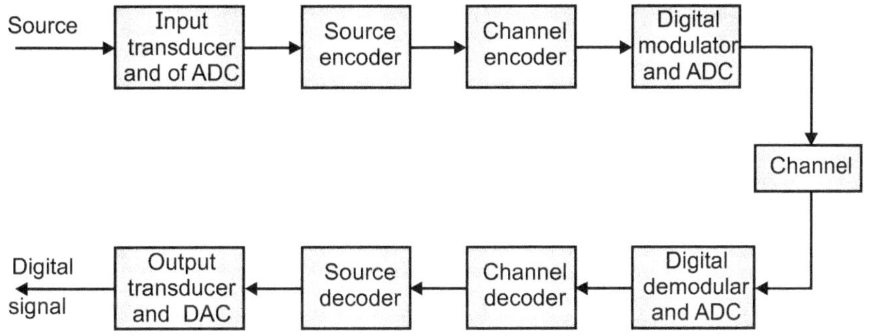


**Fig. 3.1: Digital communication system**

### 3.1.6 Advantages

1. The effect of distortion, noise and interference is much lesser in digital signals, as they are less affected.
2. Digital circuits are more reliable.
3. Digital circuits are easy to design and comparatively cheaper.
4. The hardware implementation in digital circuits, is comparatively more flexible.
5. The occurrence of cross-talk is very rare.
6. The signal is unaltered, as the pulse needs a high disturbance to alter its properties, which is very difficult.
7. Signal processing functions, such as encryption and compression are employed in digital circuits to maintain the secrecy of the information.
8. The probability of error occurrence is reduced by employing error detecting and error correcting codes.
9. Spread spectrum technique is used to avoid signal jamming.
10. Combining digital signals using Time Division Multiplexing (TDM) is easier than combining analog signals using Frequency Division Multiplexing (FDM).
11. The configuring process of digital signals is comparatively easier.
12. Digital signals can be saved and retrieved more conveniently.
13. Many of the digital circuits have almost common encoding techniques and hence similar devices can be used for a number of purposes.
14. The capacity of the channel is effectively utilized by digital signals.

### 3.1.7 Applications

1. Digital audio transmission
2. Digital audio recording
3. Telephone channels
4. Improve fidelity
5. Low-pass filter
6. Samplization
7. Quantization
8. Satellite communications
9. Telecommunications
10. Fibre optics
11. Aerospace and defence
12. Security and surveillance
13. Emergency health care
14. Video consulting
15. Distance learning
16. Medical education
17. Transfer of medical data
18. Medical care delivery, diagnostics
19. Medical consultation
20. Medical treatment at a distance
21. Nursing homes or retirement centres
22. Home monitoring for patients
23. Health care in disasters
24. Image processing for pattern
25. Digital Signal Processing (DSP)
26. Pattern recognition
27. Robotic vision
28. Image enhancement
29. Missile guidance.

## 3.2 FADING CHANNEL

### 3.2.1 Introduction

- As we know wireless communication system consists of transmitter and receiver, the path from transmitter to the receiver is not smooth and the transmitted signal may go through various kinds of attenuations including path loss, multipath attenuation, etc. The signal attenuation through the path depends on various factors.
- Various factors are time, radio frequency and path or position of transmitter or receiver. The channel between transmitter and receiver can be time varying or fixed depending upon, whether the transmitter or receiver are fixed or moving with respect to each other.

### 3.2.2 Definition

- **In wireless communications, fading is a variation of the attenuation of a signal with various variables.**
- These variables include time, geographical position, and radio frequency. Fading is often modelled as a *random process.*
- **A fading channel is a communication channel that experiences fading.**
- In wireless systems, fading may either be due to *multipath propagation,* referred to as multipath induced fading, weather (particularly rain), or shadowing from obstacles affecting *wave propagation,* sometimes referred to as *shadow fading.*

### 3.2.3 Basic Concepts

- The presence of reflectors in the environment surrounding a transmitter and receiver create multiple paths that a transmitted signal can traverse. As a result, the receiver sees the *superposition* of multiple copies of the transmitted signal, each traversing a different path. Each signal copy will experience differences in *attenuation, delay* and *phase shift*, while traveing from the source to the receiver.
- This can result in either constructive or destructive interference, amplifying or attenuating the signal power seen out. The receiver strong destructive interference is frequently referred to a *deep fade* and may result in temporary failure of communication due to a severe drop in the channel *signal to noise ratio* (SNR).
- A common example of deep fade is the experience of stopping at a traffic light and hearing an FM broadcast degenerative into static, while the signal is re-acquired, if the vehicle moves only fraction of a meter. The loss of this broadcast is caused by the vehicle stopping at a point, where the signal experienced severe destructive interference. Cellular phones also exhibit similar momentary fades.
- Fading channel models are often used to model the effects of electromagnetic transmission of information over the air in cellular networks and broadcast communications. Fading channel models are also used in underwater acoustic communications to model the distortion caused by the water.

### 3.2.4 Explanation

- Fading can cause poor performance in a communication system, because it can result in a loss of signal power without reducing the power of the noise. This signal loss can be over some or all of the signal bandwidth. Fading can also be a problem as it changes over time; communication systems are often designed to adopt to such impairments, but the fading can change faster than the adaptions can be made.
- In such cases, the probability of experiencing a fade (and associated bit errors as the signal-to-noise ratio drops) on the channel becomes the limiting factor in the link's performance. The effects of fading can be combated by using *diversity* to transmit the signal over multiple channels that experience independent fading and coherently combining them at the receiver.
- The probability of experiencing a fade in this composite channel is then proportional to the probability that all the component channels simultaneously experience a fade, a much more unlikely event.
- *Diversity* can be achieved in time, frequency or space besides diversity techniques such as application of *cyclic prefix* e.g. in **OFDM** and *channel estimation* and *equalization* can also be used to tackle fading.

### 3.2.5 Characteristics

1. Width
2. Depth
3. Velocity
4. Path loss
5. Interference
6. Doppler shift
7. Discharge
8. Flow resistance
9. Slope

### 3.2.6 Types

- The terms *slow* and *fast* fadings refer to the rate at which the magnitude and phase change imposed by the channel on the signal changes. The *coherence time* is a measure of the minimum time required for the magnitude change of phase change of the channel to become uncorrelated from its previous value :
  1. Slow fading
  2. Fast fading
  3. Block fading
  4. Frequency selective fading
  5. Flat fading.

## 3.3 CHANNEL MODELING

### 3.3.1 Introduction

- The medium between the transmitting antenna and the receiving antenna is greatly termed as *'channel'*. In wireless transmission, the characteristics of the signal as it travels from the transmitter to the receiver. The signal characteristics are due to several phenomena as under :
  1. Existence of line-of sight of path between the antennas.
  2. Reflection, refraction and diffraction of the signal due to the objects in between the antennas.
  3. The relative motion between the transmitter and receiver, and the objects in between them.
  4. The signal attenuation as it travels through the medium.
  5. The received signal can be obtained from the transmitter signal, if we can accurately model channel in between the antennas.
- It is quite difficult to model the real world environment.
- The complex of electromagnetic propagation phenomenon involved in modern wireless communications are taken into account by appropriate *channel modeling*. Statistical channel models ar a powerful tool for communication engineers, since they are able to capture the fundamental behaviour of the *wireless channel* with reasonably simple mathematical formulation.
- New communication scenarios and services demand novel *statistical channel* models or extensions of those used in the field of wireless communications.

### 3.3.2 Definition

- **A channel is a band of radio frequency (RF) used for the wireless communication.** Radio frequency (RF) channels are an important part of wireless communication. Each IEEE wireless standards specify the channels that can be used. **A communication channel is the medium used to transmit information from one point to the other.**
- **Ideally, modeling a channel is calculating all the physical processing affecting a signal from the transmitter to the receiver.**
- **A channel modeling is an essential piece of a physical layer communication simulation. A channel modeling the impulse response of the channel medium in the time domain of its fourier transform in the frequency domain.** In general, the channel impulse of a wireless communication system varies randomly over time.

### 3.3.3 Mechanisms Involved

- Three main mechanisms of electromagnetic wave propagation in channel modeling are as under :

  1. Reflection, 2. Refraction, 3. Scattering.

### 3.3.4 Description

- In data-driven channel modulating, one has a set of collected channel measurements and is looking for a model that can find out the main statistics or features of the measurement data. With this model, new samples can be generated with the same statistics of the channel. Furthermore, there are few parameters, such as the user speed, which in some form affect the statistics of the channel.
- As for a formal problem definition, assume that many measurements of one particular channel type, i.e. a specific environment like an urban area, are given. Each channel measurement is performed at a particular user speed. So, there is a set of channel measurements "$X_{urban}$" $[X_1, X_2, \ldots, X_n]$ containing 'n' samples of the channel measurements and a set of user possible speeds $V = [V_1, V_2, \ldots, V_n]$, where n is the number of distinct user speeds. Each $X_i$ is collected at a certain speed of the user. Note that when the environment changes (channel type changes), another set of measurements like $X_{rural}$ is obtained, where its elements are drawn from different statistics, and thus one needs to build another model for that environment.
- The aim is to purpose of model, which can learn the statistical distribution of the measurement data of different channel types, and generate new samples for that environment. The set of generated samples X, should have the same distribution as the measurement data X. The model should also capture the effect of different user speeds on channel measurements and generate channels with respect to a specific desired user speed.
- For evaluating the performance of such model, a metric is needed to evaluate how the sample in X emulate the statistics of the samples in X. One should employ matrics, that can demonstrate that the samples of the model trained using data set X are statistically similar to the samples in X can evaluate the *dissimilarity* between the statistical distribution of the samples of the model trained using data set X and the samples of another set like Y, measured for another channel type user speed.

### 3.3.5 Applications

1. Multipath fading can be reduced by the selection of proper channel modeling.
2. Multiple power amplifier helps in increasing transit power of a system.
3. Accurate channel modeling helps in attaining higher throughputs.
4. Interference between transmitter and receiver can be minimized by the help of channel modeling.

## 3.4 DIGITAL SIGNALING

### 3.4.1 Introduction

- It is important to node in getting started that communication system designed to carry information in digital forward communication systems designed to carry information to analog forms are both *analog system.* It is the information format that is digital or analog and not the signal transmitted over the wireless link.
- Conventionally, **system that carry information in digital format are known as digital communication systems.** These systems differ from analog systems in the design goal of the physical communication link. Analog links are designed with the object of ensuring that the user at the receiving end of the link is provided with a faithful replica of the information.
- Preserving the waveform is not important, unless doing so contributes this goal. The distinction between preserving waveform and preserving information will become clear, as we instigate a simple base band digital link.
- We will develop the basic indicator of performance equality, the bytes error rate, and we will show how performance depends on signal design, as well as other factors not entirely under the designer's control. We will investigate optimal receiver designer for the *Additive White Gaussian Noise* (AWGN) channel.
- Once we have presented the base band digital link, we will add modulation, which provides a number of advantages, not the least of which is that it enables radio transmission. To implement a data link based on a modulated carrier, a link designer must choose, both symbol and modulation formats.

- The choice of format has implementations complexity, power efficiency, adjacent channel interference, and flexibility that the designer must take into account. In most cases, a systems engineer works within the constraints of a link budget, cost and complexity limits, spectrum regulations and packing and power concerns to select the proper signaling format.
- Spread spectrum signaling is introduced. It is a modulation technique that broadens the bandwidth of the transmitted signal in a manner unrelated to the information to be transmitted. It was originally developed to provide cancelment and security, but was subsequently found to be very effective in making signals resilient in the presence of interference and frequency selective fading.

### 3.4.2 Definition

- **Digital signal is a signal that is being used to represent data as a sequence of discrete values at any given time, it can only take on one of a finite number of values.**
- This contrasts with an *analog signal*, which represents *continuous* values at any given time, it represents a *real number* a continuous range of values.
- In digital electronics, **a digital signal is a pulse train (a pulse amplitude modulated signal), i.e. a sequence of fixed width square wave electrical power or light pulses, each occupying one of a discrete number of levels of amplitude.**
- The pulse trains in *digital circuits* are typically generated by MOSFET devices due to their rapid on-off *electronic switching* speed and LSI capability.
- **In digital signal processing, a digital signal is a representation of a physical signal that is sampled and quantized**.
- A digital signal is an obstruction, which is discrete in time amplitude. **The digital signal is sequence of codes drown from a finite set of values.** The digital may be stored, processed or transmitted physically as a Pulse Coded Modulation (PCM) signal.
- **In digital communication, a digital signal is continuous - time physical signal, alternating between a discrete number of waveforms representing a pitot room.** Digital signals consist of rapid transitions between two voltage levels that are labelled as '0' and '1'.
- **A digital signal is one in which a change in voltage and the time at which it occurs, are of very much more importance than the precise size of the charge or the exact shape of the waveform.**

### 3.4.3 Limitations of Analog Signal

1. It works on continuous data.
2. It is used to measure natural or physical values.
3. It is specific to a particular task. Hence, it is not versatile.
4. Its accuracy is not low.
5. Its output form is like curve, line or graph.

### 3.4.4 Characteristics of Digital Signal

1. Adaptibility
2. Continuity
3. Representation
4. Data types
5. Signal type
6. Medium of transmission
7. Type of values
8. Security
9. Bandwidth
10. Hardwire
11. Data storage
12. Portability
13. Data transmission
14. Impedance
15. Power consumption
16. Recording of data
17. Use
18. Rate of data transmission.

### 3.4.5 Advantages

1. It is more secure and does not get damaged by noise.
2. It uses low bandwidth.
3. It allowed the signals to be transmitted over a length distance.
4. It has higher rate transmission.
5. We can translate the messages, audio, video into device language.
6. It enables the transmission of multidirectional concurrently.

### 3.4.6 Digital Signaling on Frequency Non-selective (Flat) Fading Channels

- The performance of a digital modulation scheme is degraded by many transmission impairments including fading, delay spread, Doppler spread, co-channel and adjacent channel interference, and noise.
- Fading causes a very low instantaneous received Signal-to-Noise Ratio (SNR) or Carrier-to-Noise Ratio (CNR), when the channel exhibits a deep fade, delay spread causes Inter Symbol Interference (ISI) between the transmitted symbols, and a larger Doppler spread is indicative of rapid channel variation and necessitates a receiver with a flat convergent algorithm.
- Co-channel interference, adjacent channel interference, and noise, are all additive distortions that degrade the bit error rate performance by reducing the CNR or SNR.
- We first introduce a vector representation for digital signaling on a flat fading channels with AN GN. Later a generalized analysis provided for the error rate performance of digital signaling on flat fading channels. We then derive the structure of the optimum coherent receiver for the detection of known signals in AWGN.
- The error probability performance of various coherently detected digital signaling schemes is considered including Phase-Shift Keying (PSK), Quadrature Amplitude Shift Keying (QASK), orthogonal signals and Orthogonal Frequency Division Multiplexing (OFDM).
- Other types of detection schemes include differential detection of differentially encoded binary phase shift keying and differentially encoded $\pi/4$-phase shifted quadrature phase shift keying. We also consider non-coherent detection of orthogonal signals. Finally, we consider coherent and non-coherent detection of continuous phase modulated signals.

## 3.5 DIVERSITY BRANCHES

### 3.5.1 Introduction

- One of the most powerful techniques to mitigate the effect of fading is to use diversity combining of independently fading signal paths. Diversity combining uses the fact that independent signal paths have a low probability of experiencing deep fades simultaneously.
- The idea behind diversity is to send the same data over independent fading paths. These independent paths are combined in some way such that the fading of the resultant signal is reduced. Diversity techniques that mitigate the effect of multipath fading are called *micro-diversity*. Diversity to mitigate the effects of shadowing from building and objects is called *macro-diversity*.

### 3.5.2 Definition

- **In telecommunications, a diversity technique is a method for improving the reliability of a message signal by using two or more combination channels with different characteristics.**
- Diversity is mainly used in *radio communication* and is a common technique for combating fading and *co-channel interference* and avoiding error bursts. It is based on the fact that individual channels experience different levels of *fading* and *interference*.
- Diversity is the technique used to compensate using two or more receiving antennas.

### 3.5.3 Basic Concept

- Transmit the signal via several independent diversity branches to get independent signal replicas. In other words, to have diversity, we need multiple branches independent fading and process branches to reduce fading probability.

### 3.5.4 Techniques

- The following classes of diversity schemes can be identified as under :
  1. Time diversity
  2. Frequency diversity
  3. Space diversity
  4. Polarization diversity
  5. Multi-user diversity
  6. Co-operative diversity
- It is implemented by using two or more receiving antennas.

### 3.5.5 Requirements

1. Multiple branches.
2. Low correlation between branches.

### 3.5.6 Classification

- The classification of fading channels depends on the properties of transmitted signal. The two-way classification based on time delay spread and Doppler spread gives rise to four different types of channels:
  1. Flat slow fading
  2. Flat fast fading
  3. Frequency selective slow fading.
  4. Frequency selective fast fading.

### 3.5.7 Realization of Independent Fading Paths

- There are many ways of achieving independent fading paths in a wireless system. One method is to use *multiple transmit* or *receiver antennas*, also called an *antenna array*, where the elements of arrays are separated in distance. This type of diversity is referred to a *space diversity.*
- Note that with receiver, independent fading paths are realized without an increase in transmit signal power or bandwidth. *Coherent combining* of the diversity signals leads to an increase in SNR at the receiver over the SNR that would be obtained with just a single *receive antenna.*
- To obtain independent paths through transmitter space diversity, the *transmit power* must be divided analog multiple antennas. Space diversity also requires that the separation between antennas be such that the fading amplitudes corresponding to each antenna are approximately independent. A second method of achieving diversity is by using either two *transmit antennas* or two *receive antennas* with different *polarization.*

### 3.5.8 Disadvantages of Polarization

- There are two disadvantages of polarization :
  1. You can have at most two diversity branches, corresponding to the two types of polarization.
  2. Polarization diversity losses effectively half the power (3 dB), since the transmit or receive power is divided between two differently polarized antennas.

## 3.6 DIVERSITY COMBINING

### 3.6.1 Definition

- **Diversity combining is the technique applied to combine the multiple received signals of a diversity reception device into a single improved signal.**

### 3.6.2 Various Techniques

- Various diversity combining techniques can be distinguished as follows :
  1. Equal-gain combining
  2. Maximal ratio combining
  3. Switched combining
  4. Selection combining

### 3.6.3 Equal Gain Combining

- All the received signals are summed coherently.
- On the $i^{th}$ receive antenna, equalization is performed at the receiver by dividing the received symbol Y : by the apriori known phase of $h_i$. The channel $h_i$ is represented in polar form as $|hi|e^{(i)}$. The decoded symbol is the sum of the phase compensated channel from all the received antennas.

- For PSK modulation schemes, the equalization by the phase of the channel coefficients suffice. However for QAM case, we need to compensate for the amplitude also, when equalizing.

### 3.6.4 Maximal-Ratio Combining

- Maximal-ratio combining is used in large phased-array systems. The received signals are weighted with respect to their SNR and then summed. The resulting SNR yields $\sum_{k=1}^{N} SNR_k$, where $SNR_k$ is the SNR of the received signal k.
- Sometimes more than one combining technique is used. For example, **lucky imaging** uses selection combining to choose typically the best 10% images, followed by equal-gain combining of the selected images.
- Maximal Ratio Combining (MRC) is the most complex scheme in which all branches are optimally combined at the receiver. It requires scaling and cophasing of individual branches.
- All the signals of M branches are weighted according to their individual signal voltage to noise power ratios and then summed. Thus, MRC produces as output SNR, which is equal to the sum of the individual SNRs.

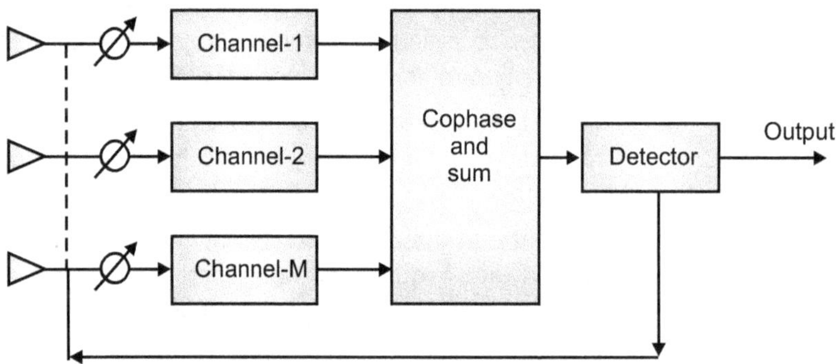


**Fig. 3.2 : Maximal ratio combining**

- The advantage of MRC is providing an output with an acceptable SNR even when none of the individual signal are themselves acceptable. Best statistical reduction of fading is achieved by this method.

### 3.6.5 Switched Combining

- The receiver switches to another signal, when the currently selected signal drops below a predefined threshold. This is also often called **'scanning combining'**.
- When the signal quality of the used branch is good, there is no need to look for (to use) other branches. Other branches are needed only when the signal quality deteriorates. Following two categories can be used :
  1. Switch-and-examine strategy.
  2. Switch-and-stay strategy.
- Switching between branches will introduce discontinuities in the combined signal.

### 3.6.6 Selection Combining

- Of the N received signals, the stronger signal is selected. When the N signals are independent and **Rayleigh distributed**, the expected diversity gain has been shown to be $\sum_{k=1}^{N} \frac{1}{k}$, expressed as a power ratio.
- Therefore, any additional gain diminishes rapidly with the increasing number of channels. This is a more efficient technique than switched combining.

- Other signal combination techniques have been designed for noise reduction and have found applications in single molecule **biophysics, chemometrics** among other disciplines.
- Selection combining is based on the principle of selection of the best signal among all the signals received from different branches at the receiving end. In this method, the receiver monitors the SNR of the incoming signal using switch logic.
- If there are M independent Rayleigh fading channels providing M diversity branches, gain of each branch is adjusted to provide equal average SNR for each branch, cophasing is not required. The branch with highest instantaneous SNR is connected to demodulator. Instantaneous selection is not possible.

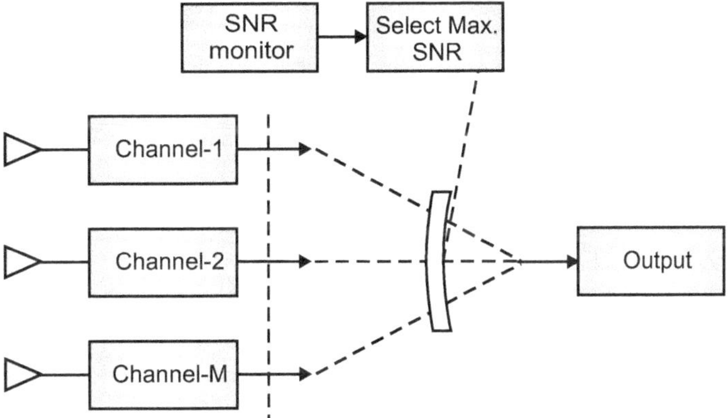


**Fig. 3.3 : Selection combining**

- Therefore, practical diversity selection system should be designed with small constant of the selection circuit than the fading rate.

## 3.7 MULTIPATH FADING

### 3.7.1 Basic Concept

- Multipath fading is a feature that needs to be taken into account, when designing or developing a radio communications system in any terrestrial radio communications system. The signal will reach the receiver not only via the path, but also as a result of reflections from objects such as buildings, hills, ground, water etc. that are adjacent to the main path.
- The overall signal at the radio receiver is a summation of the variety of signals being received. As they all have different path lengths, the signal will add and subtract from the total dependent upon their relative phases. At times, there will be changes in the relative path lengths. This could result from either the radio transmitter or receiver moving or any of the objects that provides a refractive surface moving.
- This will result in the phases of the signals arriving at the receiver changing, and in turn this will result in the signal strength varying as a result of the different way in which the signals will sum together. It is this that causes the fading that is present on many signals.

### 3.7.2 Explanation

- Multipath fading may also cause distortion to the radio signal. As the various paths that can be taken by the signals vary in length, the signal transmitted at a particular distance will arrive at the receiver over a spread of times. This can cause problems with the phase distortion and inter-symbol interference, when data transmissions are made. As a result, it may be necessary to incorporate features within the radio communication system that enables the effects of these problems to be minimum.

### 3.7.3 Flat Fading

- This form of multipath fading affects all the frequencies across a given channel either or almost equally. When flat multipath fading is experienced, the signal will just change in amplitude, rising and falling over a period of time; with movement from one position to another.

### 3.7.4 Selective Fading

- Selective fading occurs, when the multipath fading affects different frequencies across the channel to different degrees. It means that the phase and amplitudes of the signal will vary across the channel.
- Sometimes, relatively deep nulls may be experienced and this can give rise to some reception problems. Simply maintaining the overall amplitude of the receiving signal will not overcome the effects of selective fading and some form of equalization may be needed.
- Some digital signal formats, e.g. OFDM are able to spread the data over a wide channel so that only portion of the data is lost by any nulls. This can be reconstituted using forward error correction techniques and in this way, it can mitigate the effects of selective multipath fading.

**Practice Questions**

1. What is digital communication ?
2. Why do you need digitization ?
3. What is fading ? What is fading channel ?
4. Explain basic concept of fading channel.
5. List various types of fading.
6. State important characteristics of fading channel.
7. What is channel modeling ?
8. Describe channel modeling.
9. List important applications of channel modeling.
10. What is digital signaling ?
11. What are the limitations of analog signal ?
12. What are the important characteristics of digital signaling ?
13. State advantages of digital signaling.
14. Describe the digital signaling over a frequency non-selective slowly fading channel.
15. What is diversity branching ? State basic concept of diversity branching.
16. List various diversity techniques.
17. Describe realization of independent fading paths.
18. What is diversity combining ?
19. List various techniques of diversity combining.
20. Define the following terms :
    (a) Equal gain combining
    (b) Maximum-ratio combining
    (c) Switched combining, and
    (d) Selection combining.
21. Describe multipath fading concept.

✍✍✍

# MULTIPLE ACCESS TECHNIQUES FOR WIRELESS COMMUNICATIONS

**Syllabus**

Introduction, Frequency Division Multiple Access (FDMA), Time Division Multiple Access (TDMA), Spread Spectrum Multiple Access, Space Division Multiple Access, Packet Radio Protocols; Pure ALOHA, Slotted ALOHA.

## 4.1 MULTIPLE ACCESS TECHNIQUES

### 4.1.1 Introduction

- In wireless communication systems, it is often desirable to allow the subscriber to send information simultaneously from the Mobile Station (MS) to the Base Station (BS), while receiving information from the base station to the mobile station.
- A cellular system divides any given area into cells, where a mobile unit in each cell communicates with a base station. The main aim in the cellular system design is to be able to increase the capacity of the channel, i.e., to handle as many cells as possible in a given bandwidth with a sufficient level of quality of service.
- In telecommunications and computer networks, a *multiple access technique* allows more than two terminals connected to the same transmission medium to transmit over it and to share its capacity. Examples of shared physical media are wireless networks, bus networks and point-to-point links operating in *half duplex mode*.
- A multiple access technique is based on multiplexing, that allows several data streams or signals to share the same communication channel or transmission medium.

### 4.1.2 Definitions

- **In telecommunications and computer networks, multiple access is a technique that allows more than two terminals connected to the same transmission medium to transmit over it and to share its capacity.** It is also called channel access technique.
- *Multiple access is a technique that lets multiple mobile user share the allotted spectrum in the most effective manner.*
- **Multiple access technique is a multiplexing technique that provides communication services to multiple user in a single bandwidth wired in or wireless medium.**

### 4.1.3 Types

1. Frequency Division Multiple Access (FDMA).
2. Time Division Multiple Access (TDMA).
3. Code Division Multiple Access (CDMA) or Spread-Spectrum Multiple Access (SSMA).
4. Space Division Multiple Access (SDMA).
5. Power Division Multiple Access (PDMA).

## 4.2 FREQUENCY DIVISION MULTIPLE ACCESS (FDMA)

### 4.2.1 Introduction

- Frequency Division Multiple Access (FDMA) is one of the most common analog multiple access techniques. The frequency band is divided into channels of equal bandwidth so that each conversation is carried on a different frequency.
- The best example of this is the cable television system. The medium is a single coaxiable cable that is used to broadcast hundreds of channels of video or audio programming to homes. The coaxial cable has a useful bandwidth from about 4 MHz to 1 GHz. This bandwidth is divided into 6 MHz channels.
- One of the older FDMA systems is the original analog telephone system, which used a hierarchy of frequency multiplex technique to put multiple telephone calls on single line.

### 4.2.2 Definition

- **Frequency Division Multiple Access (FDMA) is a channel access technique used in some multiple access protocols.**
- *Frequency division multiple access* (FDMA) *is the division of the frequency band allocated for wireless cellular telephone communication into 30 channels, each of which can carry a voice conversation or, with digital service carry digital data.*
- FDMA is the process of dividing one channel or bandwidth into multiple individual bends, each for use by a single user.
- Each individual band or channel is wide enough to accommodate the signal spectra of the transmission to be propagated. The data to be transmitted is modulated onto each subcarrier, and all of them are linearly mixed together.

### 4.2.3 Description

- In FDMA technique, guard bands are used between the adjacent signal spectra to minimize cross-talk between the channels. A specific frequency band is given to one person, and it is received by identifying each of the frequency in the first operation of analog mobile phone.
- FDMA systems use low bit rates (large symbol times) as compared to average delay spread.

### 4.2.4 Advantages

1. It reduces the bit rate information.
2. The use of efficient numerical codes increases the capacity.
3. It reduces cost and lowers inter-symbol interference (ISI).
4. It does not need equalization.
5. It can be easily implemented.
6. It requires less number of bits for synchronization and framing.

### 4.2.5 Disadvantages

1. It does not differ significantly from analog systems.
2. Its capacity improvement depends on the Signal-to-Interference Reduction (STB) or a Signal-to-Noise Ratio (SNR).
3. Its maximum flow rate for channel is fixed and small.
4. Its guard band leads to a waste of capacity.
5. Narrow band filters cannot be realized in VLSI and therefore increases the cost.

## 4.3 TIME DIVISION MULTIPLE ACCESS (TDMA)

### 4.3.1 Introduction

- Time Division Multiple Access (TDMA) is a digital cellular telephone communications technology. It facilitates many users to share the same frequency without interference. Its technology divides a signal into different time slots, and increases the data carrying capacity.

### 4.3.2 Definition

- **Time Division Multiple Access (TDMA) is a multiple access technique for shared medium networks that allow several users to share the same frequency channel by dividing the signal into different time slots.**
- The users transmit in rapid succession, one after the other, each being its own time slot.
- **TDMA is a digital cellular telephone communication technology, which facilitates many users to share the same frequency without interference.**
- TDMA is a multiple access technique, which is widely used in VSAT and broadband satellite systems and the GSM cellular mobile systems.
- TDMA was first specified as a standard in EIA/TIA interim standard 54 (IS-541). IS-136, an evolved version of IS-54 is the united states standard for TDMA for both the cellular (850 MHz) and personal communications services (1.9 GHz) spectrum. TDMA is also used for Digital Enhanced Cordless Telecommunications (DECT).

### 4.3.3 Description

- TDMA is a complex technology, because it requires an accurate synchronization between the transmitter and the receiver. It is used in digital mobile radio systems. The individual mobile stations cyclically assign a frequency for the exclusive use of a time interval.
- In most of the cases, the entire system bandwidth for an interval of time is not assigned to a station. However, the frequency of the system is divided into sub-bands, and it is used for multiple access in each sub-band. Sub-bands are known as carrier frequencies. The mobile system that uses the technique is referred to a multi-carrier system.
- The frequency band is shared by number of users. So, peak power becomes a problem and larger by the burst communication.

### 4.3.4 Advantages

1. It permits flexible rates. So, several slots can be assigned to a user.
2. It can withstand gusty or variable bit rate traffic. So, number of slots allocated to a user can be changed frame by frame.
3. It does not require guard band for wideband system.
4. It does not require narrow band for the wide band system.

### 4.3.5 Disadvantages

1. It requires complex equalization for high data rates for broadband systems.
2. A large number of additional bits are required for synchronization and supervision due to the burst mode.
3. It needs call time in each slot to accommodate time to inaccuracies due to clock instability.
4. It increases energy consumption due to its electronics operation at high bit rates.
5. It requires complex signal processing to synchronize within short slots.

## 4.4 SPREAD SPECTRUM MULTIPLE ACCESS (SSMA)

### 4.4.1 Introduction

- Spread Spectrum Multiple Access (SSMA) is a sort of multiplexing that facilitates various signals to occupy a single transmission channel. It optimizes the use of available bandwidth. The technology is commonly used in ultra-high frequency (UHF), cellular telephone systems, bands ranging between 800 MHz and 1.9 GHz. It is also called Code Division Multiple Access (CDMA).

### 4.4.2 Definition

- **In telecommunications and radio communication, spread spectrum multiple access (SSMA) is a technique by which a signal (e.g. in electrical, electromagnetic, or acoustic signal) generated with a particular bandwidth is deliberately spread in the frequency domain, resulting in a signal with a wider bandwidth.**
- This technique is used for a variety of reasons, including the establishment of secure communications, increasing resistance to natural interference, noise, and jamming to prevent detection to limit power flux density (e.g. in satellite downlink line) and to enable multiple-access communications.
- **SSMA (CDMA) is a technique of multiple access, where several transmitters use a single channel to send information simultaneously.**

### 4.4.3 Features

1. Every user uses the full available spectrum instead of getting allotted by separate frequency.
2. It is much recommended for voice and data communications.
3. The users having the same code can communicate with each other.
4. It offers more air-space capacity than TDMA.
5. It handles very well the hand-off between base stations.

### 4.4.4 Description

- CDMA/SSMA is very different from time and frequency multiplexing. In this system, a user has an access to the whole bandwidth for the entire duration. The basic principle is that different CDMA (SSMA) codes are used to distinguish among the different users.
- Techniques generally used are direct sequence spread spectrum modulation (DS-CDMA), frequency hopping or mixed CDMA detection (JPCDMA). Here, a signal is generated which extends over a wide bandwidth. A code called *spreading code* is used to select a signal with a given code in the presence of many other signals with different orthogonal codes.

### 4.4.5 Working

- CDMA (SSMA) shows upto 61 concurrent users in a 1.2288 MHz channel by processing each voice packet with two P-N codes. There are 64 *Walsh codes* available to differentiate between calls and theoretical limits. Operational limits and quality issues will reduce the maximum number of calls somewhat lower than this value.
- In fact, many different '*signals*' passband with different spreading codes can be modulated on the same carrier to allow many different users to be supported using different orthogonal codes, interferences between the signals are minimal. Conversely, when signals are received from several mobile stations, the base station is capable of isolating each as they have different orthogonal spreading codes.
- During propagation, we mix the signals of all users, but by that you uses the same code as the code that was used at the time of sending the receiving side.

### 4.4.6 Capacity Factors

- The factors deciding the CDMA (SSMA) capacity are as under :
  1. Processing gain
  2. Signal-to-noise ratio (SNR)
  3. Voice activity factor
  4. Frequency range efficiency
- Capacity in CDMA (SSMA) is soft, CDMA has all users on each frequency and users are separated by code. This means CDMA operate in the presence of interference and noises. In addition, neighbouring cells use same frequencies, which means no re-use. So, CDMA capacity calculations should be very simple. No code channel in a cell, multiplied by no cell. But it is not that simple. Although not available code channels are 64, it may not be possible to use a single time since the CDMA frequency is the same.

### 4.4.7 Advantages

1. It requires a tight power control, as it suffers from near-far effect. All signals must have more or less equal power at the receiver.
2. It can use rake receiver to improve signal reception.
3. It may use flexible transfer.
4. Transmission bursts reduce interference.

### 4.4.8 Disadvantages

1. The code length must be carefully selected, because large code length can induce delay or may cause interference.
2. It requires time synchronization.
3. Gradual transfer increases the use of radio resources may reduce its capacity.
4. The needs of constant tight power control results in several handovers.

## 4.5 SPACE DIVISION MULTIPLE ACCESS (SDMA)

### 4.5.1 Introduction

- In traditional mobile cellular network systems, the base station has no information on the position of the mobile units within the cell and radiates the signal in all directions. Within the call in order to provide radio coverage. This method results in wasting power on transmission, when there are no mobile units to reach in addition to causing *interference* for adjacent calls using the same frequency, so called *co-channel calls.*
- Likewise, in reception, the *antenna* receives signals coming from all directions including noise and interference signals. By using *smart antenna* technology and differing spatial locations of mobile units within the cell. Space Division Multiple Access (SDMA) techniques offer attractive performance enhancements.
- The *radiation pattern* of the base station, within transmission and reception, is adapt to each user to obtain highest gain in the direction of that user. This is often done using *phase array* techniques.

### 4.5.2 Definition

- **Space or Spatial Division Multiple Access (SDMA) is technique, which is MIMO (Multiple-Input, Multiple-Output) architecture and used mostly in wireless and satellite communications.**
- *SDMA is a multiple access technique based on certain parallel spatial pipes (for used signal beams) using advanced antenna technology next to higher capacity pipes through spatial multiplexing and/or diversity, by which it is able to alter superior performance in radio multiple access.*
- **A SDMA is a multiple access technique based on creating parallel spatial pipes (focused signal beams) using advanced antenna technology next to higher capacity pipes through spatial multiplexing and/or diversity, by which it is able to after superior performance in radio multiple access communication systems, whose multiple users may need to use the communication media simultaneously.**

### 4.5.3 Features

1. All users can communicate at the same time using the same channel.
2. It is completely free from interference.
3. A single satellite can communicate with more satellites of the same frequency.
4. The directional spot-beak antennas are used and hence the base station can track a moving user.
5. It controls the radiated energy for each user in space.

### 4.5.4 Description

- Wireless transmitter or receiver units are assigned to distinct (frequency) channels and thereby allowed to communicate simultaneously. These channels can be frequency channels, time-slots in particular frequency channels, or code channels in a particular frequency channels.
- A multi-channel receiver exploits the fact that they are on different frequency channels to correctly separate the signals, which are then subsequently demodulated and passes along to the rest of the network. A multi-channel transmitter transmits signal to the wireless units in another set of distinct frequencies.
- For example, in current cellular mobile communication systems, mobile units receive transmission from base stations in channels 45 MHz above those frequency channels, they transmit information to the base antenna. This allows for simultaneous transmission and reception of information as both the base stations and mobile links.
- It is also possible to use time-multiplexing and voice compression to transmit and receive on the same frequency, however, this concept would not be amenable to high-spread continuous data transmission.

## 4.6 PACKET RADIO PROTOCOLS

### 4.6.1 Introduction

- In 1973, the Defence Advanced Research Projects Agency (DARPA) initiated research on the feasibility of using packet switched, store-and-forward radio communications to provide reliable computer communications. This development was motivated by the need to provide computer network access to mobile hosts and terminals, and to provide computer communications in a mobile environment.
- Packet radio networking offers a highly efficient way of using a multiple access channel, particularly with bursty traffic. The DADPA Packet Radio Network (PRNE) has evolved through the years to be a robust, reliable, operational experimental network. The development process has been of an incremental, evolutionary nature; as algorithms were designed and implemented, new versions of the PRNET with increased capabilities were demonstrated.
- The PRNET has been in daily operation for experimental purposes for nearly 10 years.

### 4.6.2 Definition

- **Packet Radio (PR) is a digital radio communications mode used to send packets of data.**
- Packet radio uses *packet switching* to transmit *diagrams*. This is very similar to how packets how packets of data are transferred between nodes on the *internet*. Packet radios can be used to transmit data to long distances.

### 4.6.3 Description

1. **Broadcast Radio :**

- The PRNET provides, via a common radio channel, the exchange of data between computers that are geographically separated. As a communications medium, broadcast radio provides important advantages to the user of the network. One of the benefits is mobility, a Packet Radio (PR) can operate while in motion.
- Second, the network can be installed or developed quickly; there are no wires to set up. Third advantage is the case of reconfiguration and redeployment. The PRNET protocols take advantage of broadcasting and common - channel properties to allow the PRNET to be expanded or contracted automatically and dynamically.
- A group of packet radios leaving the original area simply departs. Having done so it can function as an autonomous group and may later dejoin the original network or join another group. The broadcasting and common channel properties of radio have disadvantages too. These properties, for all practical purpose, prohibit the building of a radio that is able to transmit and receive at the same time.

- Therefore, the PRNET protocols must attempt to schedule each transmission, when the intended protocol radio (PR) is not itself transmitting. Also, transmissions often reach unintended protocol radios and interfere with intended receptions. Therefore, the protocol must attempt to schedule each transmission, when the intended protocol radio is not receiving another protocol's transmission.

2. **Automated Network Management :**

- The PRNET features fully automated network management. It is a self-configuring upon network initialization, reconfigures upon gain or loss of packet radios and has dynamic routing. The network operator simply has to ensure that the packet radios are deployed so that every packet radio is situated so as to have at least one other packet radio within line-of-sight, and then turn the radios on; the packet radio is intended to operate unattended. Once installed, the system discovers the radio connectivity between packet radios and organizers routing strategies dynamically on the basis of this connectivity.
- After initialization, more communication networks maintain a static topology. A unique feature of the PRNET is the ease with which network topology can be altered without affecting the user's ability to communicate. Although RF connectivity is difficult to predict and may abruptly change in unexpected ways as mobile packet radios move about, the automated network management proceed users used in the PRNET are capable of sensing the existing connectivity in real-time and then exploiting this connectivity in order to continuously transport data and control packets, all in a way that is totally transparent to the users.

### 4.6.4 Advantages

1. It allows simultaneous use of any conventional (frequency, time slot, or code) channel by multiplex user.
2. It mitigates hand-off and signal management problems by tracking of mobile services.
3. It enables location-related services.
4. It is independent of particular signal modulation, which provides more compatibility with current and future modulation scheme.
5. It has improved signal quality at both transmitters and receivers.
6. It has a certain amount of security by transmitting signals only in preferred directions.
7. It allows a decrease in transmitting power to be affected at the base stations.
8. It has decrease in signal degradation due to co-channel interference.
9. It provides capability for the new PCS systems.

## 4.7 COMPARISON OF CDMA AND FDMA

**Table 4.1**

| CDMA (SSMA) | FDMA |
|---|---|
| 1. The same frequency is used by each user. | 1. The same frequency is not used by each user. |
| 2. It uses coherent. | 2. There is no code word. |
| 3. Equalization is needed. | 3. Little or no equalization is needed. |
| 4. Each user has a separate code. | 4. For broadcasting, time symbols are suitable analogue links. |
| 5. It is mandatory for the receivers to know about the issues code word. | 5. Synchronization bits are not needed. |

## 4.8 ALOHA

### 4.8.1 Brief History

- One of the early computer networking designs, development of the ALOHA network was began in September 1968 at university of Hawaii under the leadership of *Norman Abramson* along with *Thomas Gaarder, Franklin Kuo, Shulin, Wasely Peterson* and *Edward Weldon*. The goal was to use low-cost commercial radio equipment to connect users on *Oahu* and other Hawaiin instants with a central time sharing computer on the main *Oahu* campus.
- In the 1970s, ALOHA random access was employed in the nascent Ethernet cable based network and then in the Marisett (now Inmorsat) satellite network.
- ALOHA channels were used in limited way to the 1980s in first generation (FG) mobile phones for signaling and control purposes. In the late 1980s, the European Digital Mobile Communication System.
- GSM greatly expanded the use of ALOHA channels for access to radio channels in mobile telephony. In addition SMS mass go texting was implemented in second generation (2G) mobile phones. In early 2000s, additional ALOHA channels were added to 2.5 G and 3G mobile phones with the widespread introduction of GPRS, using a slotted. ALOHA random access channel combined with a version of the Reservation ALOHA Scheme first analyzed by a grouped BBN.
- The first *packet broadcasting* unit went into operation in June, 1971. Terminals were connected to a special purpose *'terminal connection'* unit using RS-232 at 9600 bits/second. The initial purpose of the THE ALOHA SYSTEM is to provide a systematically different designer interaction for radio communications. This alternative method allows the system to determine, when and where radio communications are *'preferable'* to *wired communications*. It made practical means of communication and made accessibility of differing networks plausible.
- The original version of ALOHA used two distinct frequencies in a hub configuration, with the hub machine broadcasting packets to everyone on the *'outbound'* channel, and the various client machines sending data packets to the hub on the *'inbound'* channel. If data was received correctively at the hub, a short acknowledgement packet was sent to the client it, if an acknowledgement was not received by a client machine after a short wait time, it would automatically selected time interval.
- This acknowledgement mechanism was used to detect and correct for 'collisions' created, when two client machines both attempted to send a packet at the same time. ALOHA net's primary importance was its use of a shared medium for client transmissions. Unlike the ARPANET, where each node could only talk directly to a node at the other end of a wire or satellite circuit. In ARPANET all client nodes communicated with the hub on the same frequency.
- This means that some sort of mechanism was needed to control, who could talk at what time. The ALOHAnet solution was to allow each client to send its data without controlling, when it was sent, with an acknowledgement or retransmission scheme used to deal with collisions. This approach radically reduced the complexity of the protocol and the networking hardware, since nodes do not need to negotiate, **'who'** is allowed to speak.
- In the early 1980s, frequencies for mobile networks become available, and in 1985 frequencies, suitable for what become known as **Wi-Fi** were allocated as the US. These regulatory developments made it possible to use the ALOHA random-access techniques in both **Wi-Fi** and in mobile telephone networks.

### 4.8.2 Introduction

- ALOHAnet, also known as the ALOHA system, or simply ALOHA, was a pioneering *computer networking* system developed at the *University of Hawaii*. ALOHAnet became operational in June 1971, providing the first public demonstration of a wireless packet data network. ALOHA generally stood for Additive Links on Hawaii Area.

- The ALOHAnet used a new method of medium access (ALOHA random access) and experimental *Ultra High Frequency* (UHF) for its operation, since frequency assignments for communications to and from a computer were not available for commercial applications in the 1970s. But even before such frequencies were assigned, there were two other media available for the application of an ALOHA channel-cables and satellites.

### 4.8.3 Definition

- The acronym ALOHA stands for *Advocates of Linear Operation - Source Hawaii Association.*
- **ALOHA is a simple communication scheme in which each source (transmitter in computer network sends data, whenever there is a frame to send.**
- If the frame successfully reaches the destination (receiver), the next frame is sent. This protocol was originally developed at the University of Hawaii for use with *satellite communication* systems in the pacific.
- *ALOHA is a system for coordinating and arbitrating access to a shared communication network channel.*
- The original ALOHA system was used for ground based radio broadcasting, but this system has been implemented in satellite communication systems.
- **ALOHA in computer networks is a multiple access control protocol at the data link layer proposes how multiple terminals access the medium without interference or collision.**

### 4.8.4 Types

1. Pure ALOHA
2. Slotted ALOHA

## 4.9 PURE ALOHA

### 4.9.1 Definition

- *The ALOHA protocol that allows the station to transmit data at any time whenever they want and after transmitting the data packet, station wait for same time is called a pure ALOHA.*

### 4.9.2 Assumptions

1. All frames have the same length.
2. Stations control generates frame while transmitting or trying to transmit, i.e. if a station helps trying to a frame, it cannot be allowed to generate more frames to send.
3. The population of stations attempts to transmit, both new frames and old frames that collided, according to a *Poisson distribution.*

### 4.9.3 Description

- The version of the ALOHA protocol is now called as pure ALOHA and the one implemented in ALOHA net was quite simple. If you have data to send, then send the data. If while you are transmitting data, you receive any data from another station, has been a message collision. All transmitting stations will need to try resending later.
- Note that the first stage implies that pure ALOHA does not check whether the channel is busy before transmitting. Since, collisions can occur and data may have to be sent again, ALOHA cannot use 100% of the capacity of the communications channel.
- How long a station waits until it transmits, and the likelihood a collision occurs are interrelated, and both affect how efficiently the channel can be used. This means that the concept of *'transmit later'* is a critical aspect. The quality of the *back-off scheme'* chosen significantly influences the efficiency of the protocol, the ultimate channel capacity and the probability of its behaviour.
- To access pure ALOHA, there is a need to predict its throughput, the rate of successful transmission of frames.

### 4.9.4 Efficiency

The efficiency, $\eta = G \cdot e^{-2G}$ ... (4.1)

where, G is number of stations willing to transmit data.

Substitute $d\eta/dG = 0.$

Maximum value of efficiency occurs at $G = \frac{1}{2}$

Substituting $G = \frac{1}{2}$ in equation (4.1), we get,

$$\eta_{max} = \frac{1}{2} \cdot e^{-2} \times \frac{1}{2}$$

or

$$\eta_{max} = \frac{1}{2e} = 0.184 = 18.4\% \quad \text{... (4.2)}$$

Thus, maximum efficiency of pure ALOHA is 18.4%. It is very less due to large number of collisions.

## 4.10 SLOTTED ALOHA

### 4.10.1 Definition

- **The ALOHA protocol that allows the station to transmit the data only at the beginning of anytime-slot, the time being discrete globally synchronized, is called slotted ALOHA.**

### 4.10.2 Assumptions

1. All frames consist of exactly L bits.
2. Time is divided into slots of size L/R seconds, i.e., a slot equals the time to transmission frame.
3. Nodes start to transmit frames only at the beginning.
4. The nodes are synchronized so that each node knows, when the slot begins.
5. If two or more frames collide in a slot then all the nodes detect the collision event before the slot ends.

### 4.10.3 Description

- Slotted ALOHA is quite similar to pure ALOHA, differing only in the way transmission taking place. Slotted ALOHA was invented to improve the efficiency of pure ALOHA as chances of collision in pure ALOHA are very high. In slotted ALOHA, the time shared channel is divided into discrete intervals called **slots.** Instead of transmitting right at demand time, the sender waits for some time. This delay is specified as follows.
- The time line is divided into equal slots and then it is required that transmission should take place only at slot boundaries. To be more precise, the slotted-ALOHA makes some assumptions. According to these assumptions, the number of collisions that can possibly takes place is reduced by a hinge margin. And hence, the performance becomes much better compared to pure ALOHA, collisions may only take place with nevertheless, this is a substituted reduction.
- The slot can sent a frame only at the beginning of the slot and only one frame is sent in each slot. If any station is not able to place the frame onto the channel at the beginning of the slot i.e., it misses the time slot, then station has to wait until the beginning of the next time slot. There is still a possibility of collision, if the stations may to send at the beginning of the same time slot. It however has edge over pure ALOHA as chances of collisions are reduced to any half.

### 4.10.4 Protocol Flow Chart

- A station which has a frame ready will send it. Then wait for sometime. If it receiver the acknowledgement, then the transmission is successful. Otherwise the station uses a back-off strategy and sends the packet again. After many times, if there is no acknowledgement, then the station aborts the idea of transmission.

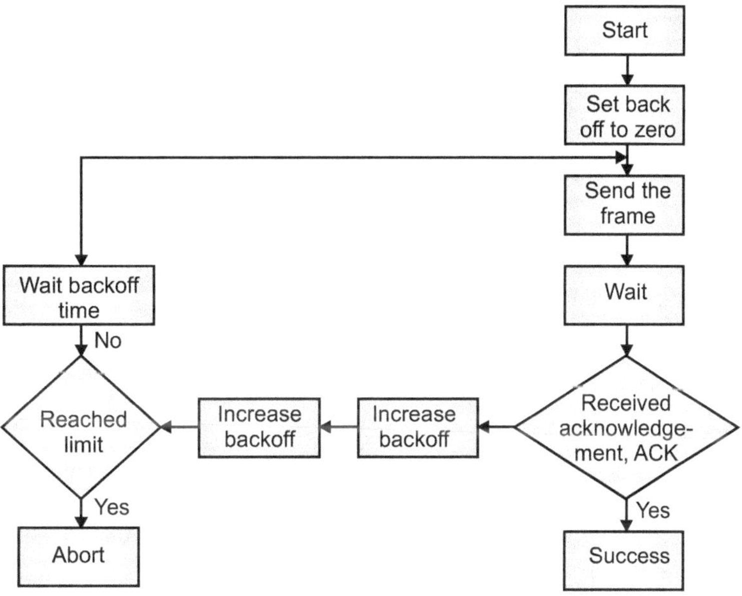


**Fig. 4.1 : Protocol flow chart for ALOHA**

### 4.10.5 Efficiency

- For maximal efficiency, $d\eta/d\theta = 0$. It occurs at G = 1.

∴ $\eta_{max} = 1 \cdot e^{-1} = 1/G$

or $\eta_{max} = 0.368 = 36.2\%$

The maximum efficiency of a slotted ALOHA is twice that of pure ALOHA and is 36.2% due to less number of collisions.

## 4.11 COMPARISON OF PURE AND SLOTTED ALOHA

**Table 4.2**

| Pure ALOHA | Slotted ALOHA |
|---|---|
| 1. Any station can transmit the data any time. | 1. Any station can transmit the data at the beginning of any time slot. |
| 2. The time is continuous and not globally synchronized. | 2. The time is discrete and globally synchronized. |
| 3. Vulnerable time in which collision may occur ($= 2T_t$). | 3. Vulnerable time in which collision may occur ($= T_t$). |
| 4. Probability of successful transmission of data packet ($= G \cdot e^{-2G}$). | 4. Probability of successful transmission of data packet ($G = G \cdot e^{4}$). |
| 5. Maximum efficiency is 18.4% only which occurs at G = 1/2. | 5. Maximum efficiency is 36.8%, which occurs at G = 1. |
| 6. The main advantage is its simplicity in implementation. | 6. The main advantage that it reduces the number of collisions and doubles the efficiency. |

**Practice Questions**

1. What is multiple access ?
2. List various types of multiple access techniques.
3. What is FDMA ?
4. List advantages and disadvantages of FDMA.

5. Explain FDMA.
6. What is TDMA ?
7. State advantages and disadvantages of TDMA.
8. What is SSMA ? List key features of SSMA.
9. Explain working principle of SSMA.
10. State capacity factor deciding the SSMA capacity.
11. List advantages and disadvantages of SSMA.
12. What is SDMA ?
13. List key features of SDMA.
14. Describe SDMA.
15. What is packet radio protocol ?
16. Describe in brief the packet radio protocols.
17. List advantages of packet radio protocols.
18. Compare CDMA and FDMA.
19. What is ALOHA ?
20. What is lower ALOHA ?
21. What are assumptions of pure ALOHA ?
22. Describe in short pure ALOHA.
23. State efficiency of pure ALOHA.
24. What is slotted ALOHA ?
25. List assumptions of slotted ALOHA.
26. Describe in short slotted ALOHA.
27. Show flow-chart of slotted ALOHA.
28. What are efficiencies of slotted ALOHA ?
29. Compare pure and slotted ALOHA.

✍✍✍

# WIRELESS SYSTEMS AND STANDARDS

**Syllabus**

AMPS and ETACS, United state digital cellular (IS-54 and IS-136), Global System for Mobile (GSM): Services, Features, System Architecture and Channel Types, Frame Structure for GSM, Speech Processing in GSM, GPRS/EDGE Specifications and Features. 3G Systems : UMTS and CDMA 2000 Standards and Specifications. CDMA Digital Standard (IS-95) : Frequency and Channel Specifications, Forward CDMA Channel, Reverse (CDMA) Channel, Wireless Cable Television.

## 5.1 WIRELESS COMMUNICATIONS

### 5.1.1 Brief History

- The term *'wireless'* has been used twice in communication history, with slight different meaning. It was initially used in 1890 for the first radio transmitting and receiving technology, as in **wireless telegraphy**, until the new word *'radio'* replaced it around 1920.
- The term wireless was revived in the 1980s and 1990s mainly to distinguish digital devices that communicate without wires. This became its primary usage in the 2000s due to the advent technologies such as *mobile broadband, Wi-Fi* and *Bluetooth.*
- *Wireless communication is the transfer of information or power between two or more points that are not connected by an electrical conductor.* The most common wireless technologies use radio waves. With radio waves, intended distances can be short, such as few maters for *Bluetooth* or as far as millions of kilometers for *deep-space radio communication.*
- It encompasses various types of fixed, mobile and portable applications including two-way radios, cellular telephones, Personal Digital Assistance (PDA) and wireless networking. Other examples of applications of *radio wireless technology* include GPS units, garage door openers, wireless computer mouse, keyboards and headsets, broadcast television and cordless telephones.
- Somewhat less common methods of achieving wireless communications include the use of other *electromagnetic* wireless technologies such as light, magnetic, or electric fields or the use of sound.
- In 1884, **Guglieno Marconi** began developing a wireless telegraph system using *radio waves*, which had been known about. Since proof of their existence in 1888 by *Heinrich Hertz,* but discounted as a communication format since they seemed, at the time, to be a short range phenomenon.
- Guglieno Marconi soon developed a *system* that was transmitting signals way behind distances any one could has predicted. *Guglieno Marconi* and *Karl Ferdinand Brown* were awarded the 1909 *Nobel Prize for Physics* for their contribution to this form of *Wireless Telegraphy. Millimeter wave* communication was first investigated by *Jagdish Chandra Bose* during 1894 to 1896, when he reached an *extremely high frequency* of upto 60 GHz in his experiments. He also introduced the use of *semiconductor*. Semiconductor junctions to detect radio waves, when he *patented* the *radio crystal detector* in 1901.
- The wireless revolution began in the 1990s, with the advance of digital *wireless networks* leading to a social revolution, and a paradigm shift from wired to wireless technology including the proliferation of commercial wireless technologies such as cell phones, mobile telephony, pagers, wireless computer networks, cellular networks, wireless internet, and laptop and handheld computers with wireless connections.

- The wireless revolution has been driven by advances in *radio frequency* (RF) and *Microwave engineering* and the transition from analog to digital RF technology, which enabled a substantial increases in *voice traffic* along with the delivery of *digital data* such as *text messaging, images* and *streaming media.*
- The MOSFET is the basic building block of modern wireless networks, including the base station modules, routers, RF circuits, radio transceivers, transmitters, and RF power amplifiers. RF CMOS is used in the radio transceivers of all modern wireless networking devices and mobile phones, and is widely used to transmit and receive wireless signals in a variety of applications, such as satellite technology e.g., GPS, Bluetooth, Wi-Fi, near field communication, mobile networks (5G and 4G), terrestrial broadcast, and automotive radar applications.
- In recent years, an important contribution to the growth of wireless *communication networks* has been interference alignment, which was discovered by *Syed Ali Jafer* at the *University of California, Irvine.* According to *Paul,* this has revolutionalized our understanding of the capacity limits of wireless networks and demonstrated the outstanding result that each user in a wireless network can access half of the spectrum without interference from other users regardless of how many users are sharing the spectrum.

### 5.1.2 Introduction

- In the present days, wireless communication system has become an essential part of various types of wireless communication devices, that permit user to communicate event from mole operated areas.
- There are many devices used for wireless communication like mobiles, cordless telephones, Zigbee, wireless technology, GPS, Wi-Fi, satellite television and wireless computer parts, current wireless phones include 3G and 4G networks, Bluetooth and Wi-Fi technologies.
- In 1897, *Guglieno Marconi* was the first to demonstrate that it was possible to establish a continuous communication system with the ship.
- The present days, wireless communication system has become an essential part of various types of communication devices that permits user.

### 5.1.3 Definition

- **Wireless communication is a data communication that transfers the information or data or power between two or more points that are not connected by an electrical conductor.** The most common wireless technologies use radio waves.
- Wireless communication incorporates all procedures and forms of connecting and communicating between two or more devices using a wireless signal through wireless technologies and devices.
- **Wireless communication is a technique of transmitting information and/or data from one point to other, without using any connection like wires, cables or any physical medium.** Generally, it comprises of a wide range of technologies, services, and applications that have come into existence to meet particular needs.
- **Wireless communication is a type of data communication that is performed and delivered wirelessly.**

### 5.1.4 Types

1. Infrared communication
2. Satellite communication
3. Broadcast radio
4. Microwave radio
5. Bluetooth technology
6. ZigBee technology
7. Wi-Fi technology
8. Mobile communication

### 5.1.5 Advantages

1. Any data or information can be transmitted faster and with a high speed.
2. Maintenance and installation is less costly for these networks.
3. The internet can be accessed from anywhere wirelessly.
4. It is very helpful for workers, doctors, etc. working on remote areas as they can be in touch with medical centres.

### 5.1.6 Disadvantages

1. An unauthorized person can easily capture the wireless signals, which spread through air.
2. It is very important to secure the wireless network, so that the information cannot be misused by unauthorized users.

### 5.1.7 Application Areas

1. Mobile telephones
2. Data communications
3. Computer peripherals
4. Energy transfer
5. Medical technologies
6. Security systems
7. TV remote control
8. Wi-Fi
9. Cell phones
10. Wireless power transfer
11. Computer interface devices
12. Wireless communication projects.

## 5.2 WIRELESS SYSTEMS AND STANDARDS

### 5.2.1 Introduction

- Wireless systems refer mainly to infrared (IB) and radio technologies. However, they may include others, where the transmission medium already exists or is shared with other services such as power lines (mains). Most of the current specifications for wireless networking are covered by the IEEE 802 standards, and more specifically, by the IEEE 802.11 wireless LAN standards - base.
- A wireless IEEE 802 network consists of one or more fixed stations. These will service multiple mobile stations. An important aspect of IEEE 802 is that mobiles can move freely between different fixed station radio ranges and still be able to maintain a seamless connection. A wireless transceiver cannot listen to the network for other transmission, while it is transmitting. This is because its own transmitter will drown out any other remote signal that may be present.
- All traffic is therefore, half-duplex in nature. Stations avoid collision by listening before transmission and by using random back-off delays to delay transmission, when the network is busy.

### 5.2.2 Definition

- **A wireless system is any collection of elements or subsystems that operate interdependently and use unguided electromagnetic waves propagation to perform some specified functions.** Each wireless system contains at least one transmitting antenna and at least one receiving antenna.

### 5.2.3 Modes of Communication

1. Radio waves
2. Free space optical
3. Sonic
4. Electromagnetic induction.

### 5.2.4 Types

- Some of the important world-wide wireless mobile radio system standards are as under :
  1. AMPS (Advanced Mobile Phone Standards).
  2. ETACS (Extended Total Access Communication System).
  3. NAMS (North American Mobile Phone System).
  4. IS-95 (Interim Standards - 95).
  5. UTMS (Universal Mobile Telecommunication System).
  6. CDMA-2000 (Code Division Multiple Access - 2000).
  7. GSM (Global System for Mobile).

## 5.3 ADVANCED MOBILE PHONE SYSTEM (AMPS)

### 5.3.1 Brief History

- The first cellular network efforts began at *Bell laboratories* and with research conducted at *Motorola.* In 1960, *John F. Mitchell* an electrical engineer oversaw the development and marketing of the first *pager* to use transistors. Motorola produces mobile telephones for automobiles, but these large and heavy models consumed too much power to allow their use without automobile engine tuning.
- *Dr. Martin Cooper,* a team member of Mitchell's team, developed the *portable cellular telephony* in 1973. At the same time Bell laboratories worked out a wireless system called Advanced Mobile Phone System (AMPS), which became the first cellular network standard in the United States in 1977-78. The world's first cellular system was implemented by Nippon Telephone and Telegraph (NTTC), Japan in 1978. The first wireless system was successfully deployed by Bell Laboratories in Chicago, Illinois in 1978.
- *Dr. Martin Cooper* produced the first commercially available *cellular phone,* small enough to be easily carried and later be introduced so called *Big phone.* In 1992, the first smart-phone, called *IBM Simon*, using AMPS was designed by Frank Canava at IBM and demonstrated in the same year at the COMDEX computer industry trade show.
- A refined version of the product was marketed to consumers in 1994 by *Bell South under* the name *Simon Personal Communication.* The *Simon* was the first device that can be properly referred to as a *smart phone.*

### 5.3.2 Introduction

- Advanced Mobile Phone System (AMPS) is a standard system based on the initial *electromagnetic radiation spectrum* allocation for cellular service by the Federal Communications Commission (FCC) in 1970. In 1983, AT & T introduced the AMPS, which became one of the most widely deployed cellular system in the united states.
- AMPS allocates frequency ranges within 800 and 900 MHz spectrum to cellular telephone. Each service provider can use half of the 824 to 849 MHz range for **receiving** signals from cellular phones and half of 869 to 894 MHz range for *transmitting* to cellular phones.
- The bands are divided into 30 kHz sub-bands, called *channels.* The receiving are called *reverse channels* and the lending (transmitting) channels are called *forward channels.* The division of the spectrum into sub-band channels is achieved by using frequency division multiple access (FDMA).
- The signals received from a transmitter cover an area called a *cell.* As a user moves out of the cell area into an adjacent cell, the user begins to pick-up the new call's signals without any noticeable transition. The signals in the adjacent cell are sent and received on different channels than the previous cell's signals to so that the signals do not interfere with each other.
- The analog service of AMPS has been updated with digital cellular service by adding to FDMA a further sub-division of each channel using Time Division Multiple Access (TDMA). This service is known as Digital AMPS (D-AMPS). Although, AMPS and D-AMPS originated for the North American cellular telephone market, they are new used worldwide with over 75 million subscribers, according to Erricson, one of the major cellular phone manufacturers.

### 5.3.3 Definition

- The long form of AMPS is **Advanced Mobile Phone System. AMPS is an analog mobile phone system standard developed by Bell Laboratories, and officially introduced, in the Americas on October 13, 1983, Israel in 1986, Australia in 1987, Singapore in 1988 and Pakistan in 1990.**
- **AMPS is a wireless standard system for analog signal cellular telephone service in the United States and is also used in other countries.**

### 5.3.4 Characteristics

1. Analog FM modulation.
2. RF bandwidth 30 kHz. The band can accommodate 832 duplex channels, among which 21 are reserved for set-up, and the rest for voice communication.
3. Frequency allocated by FCC on 824 to 849 MHz for downlink and 869 to 894 MHz for uplink traffic.
4. Uses the same system throughout the US.
5. Available in US, Canada, Hong Kong, New Zealand, Thailand etc.
6. A narrowband versus exists with a 10 kHz channel spacing, such that 2496 channels can be assigned, instead of 832 for the normal AMPS mode.

### 5.3.5 Features

1. It is an analog system based on the initial electromagnetic spectrum allocation for cellular service by the FCC.
2. It uses FDMA for multiple simultaneous conversations.
3. Frequency ranges within 800 and 900 MHz are allocated for cellular telephones in AMPS. Half of the signal is used for sending signals and half is used for receiving signals.
4. It has high bandwidth requirement particularly, when the number of conversations is very high.
5. If was the first system to use hexagonal cells so, the pioneers of AMPS had coined the term coined cellular.
6. The cells in AMPS are 10 km to 20 km across.
7. Since, it was an analog technology, it suffered from noise and eaver dropping.

### 5.3.6 Standards

- AMPS was originally standardized by American National Standard Institute (ANSI) as ETA/IIA/IS-3, which was succeeded by EIA/TIA-553 and TIA interium standard with digital technologies, the cost of wireless services so low that the problem of cloning has virtually disappeared.

### 5.3.7 Working Principle

- AMPS uses frequency in the 800 to 900 MHz range of the radio spectrum. It modulates 3 kHz voice channel onto 30 kHz FM carrier signals using FDMA to create a series of 30 kHz channels. Separate channels are used for base station to mobile transmission (forward channels) and mobile station to base transmission (backward channels).
- The resulting allocation of bandwidth for each channel results in a maximum of approximately 800 simultaneous phone conversations per operator. Because the population of most cities would suggest that 800 simultaneous phone conversations is far from enough, the idea was developed to partition the coverage of cities into a number of small areas called *'cells'*.
- Each base station uses a limited power transmitter with a directional antenna to provide coverage for a small geographical cell. A typical cell ranges from 5 km to 20 km in size, depending on whether the coverage is in a densely populated urban area or sparsely populated rural area. Mobile phone users also have limited transmission power, meaning that communication is usually limited to immediate cell the user is in.
- As a user moves from one cell to another, the signal is smoothly picked up from the new cell. Adjacent cells use different frequencies, which prevent interference. Handheld AMS cellular phones have power level generally under 0.6 watt with a range of about 5 miles (8 km) from the base station, while power levels in vehicle-mounted phones reach upto 3 watts with a range of 15 miles (24 km).
- Base stations themselves generally have power levels upto about 1 kW. Because of the need for data transmission and security (encryption), digital cellular phone services are licencing properly.

### 5.3.8 Network Layout

- The D-AMPS 800/1900 system architecture is similar to most other cellular system architectures, e.g. GSM network architecture as shown in Fig. 5.1. It consists of a switching system, an operator and support system, base stations and mobile station.

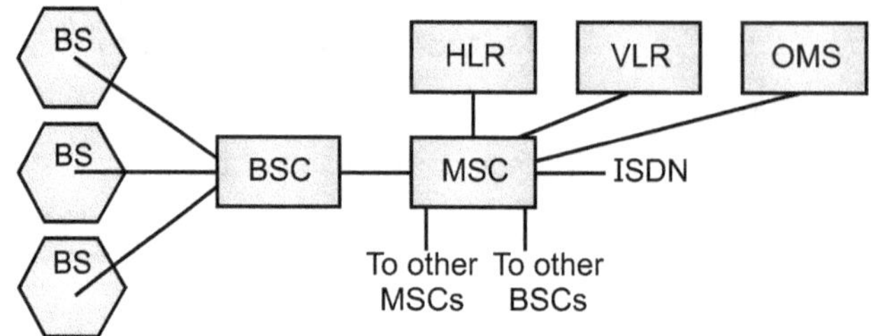


**Fig. 5.1 : AMPS architecture**

- The switching system consists of following main entities :
  1. MSC (Mobile Switching Center).
  2. Authentication Centre (AC).
  3. Base Station (BS).
  4. Operation and Support System.
  5. Mobile Station (MS).

1. **Mobile Switching Center (MSC) :**
   - The MSC performs the telephony functions for the network. It controls calls to and from other telephone and data communication network, such as Public Switched Telephone Networks (PSPN), Integrated Service Digital Networks (ISDN), Public Land Mobile Networks (PL-MN) and Public Data Networks (PDN).
   - The Visitor Location Register (VLR) data base contains all temporary subscriber information needed by the MSC to serve visiting subscribers, who are temporarily in the area of the MSC. The Home Location Resister (HLR) database stores and manages user subscriber information including their service profile, location information and activity status.
2. **Authentication Center (AC) :**
   - The Authentication Center (AC) supports authentication and encryption functionality. It verifies the user's identity by authentication and ensures the confidentiality of each call by encryption. This protects network operator against fraud. The Message Center (MC) supports message services.
3. **Base Station (BS) :**
   - The Base Station (BS) is the radio equipment needed to serve each cell in the network. One BS site may serve more than cell.
4. **Operation and Support System (OSS) :**
   - The operation and support system (SS) supports operation and maintenance activities in the network to allow for reliable and cost efficient operation.
5. **Mobile Station (MS) :**
   - The Mobile Station (MS) comprises all user equipment and software needed for communication with a mobile network. It consists of following four main components :
     (i) Mobile Termination (MT)
     (ii) Terminal Equipment (TE)
     (iii) Terminal Adapter (TA)
     (iv) Subscriber Identity Mobile (SIM).

## 5.4 EUROPEAN TRAIN CONTROL SYSTEM (ETCS)

### 5.4.1 Brief History

- The European railway network grow from separate national networks with little more in common than *standard guage*. Notable differences include *voltages, loading guage, couplings,* signaling and control systems. By the end of the 1980s, there were 14 national standard train control systems in use across the *European Unions* (EU) and the advent of high-speed trains showed that signaling based on line side signals is insufficient.

- Both factors led to efforts to reduce the time and cost of cross-border traffic. On 4 and 5 October 1989, working including Transport Ministers revolved a master plan for a trans. *European high-speed rail networks,* the first time that ETCS was suggested. The commission communicated the decision to the European Council, which approved the plan in its resolution of December 17, 1990.
- This led to a resolution on 91/440/EEC of July 28, 1991, which mandated the creation of requirements list for interoperability in high-speed rail transport. The rail manufacturing industry and rail network operators had agreed on creation of interoperability standards in June 1991.
- Until 1993, the organizational framework was created to start technical specifications that would be published as Technical Specifications for Interoperability (TSI). The mandate for TSI was resolved by 93/98/EEC. In 1995, a development plan first mentioned the creation of the European Rail Traffic Management System (ERTMS).
- Because ETCs is in many parts implemented in software, some wording from software technology is used. Versions are called System Requirements Specifications (SRS). This is a bundle of documents, which may have different version for each document. A main version is called Base Line (BL).
- The specification was written in 1996 in response to **EU Council Directive 96/48/EC 99** of July, 23, 1996 on interoperability of the trans-European high-speed rail system. First the European Railway Research Institute (ERBI) was instructed to form 6 railway operations that took over the lead role in the specification.
- The standardization went on for the next two years and it was felt to be slow for some industry patterns - 1998 saw the foundation of Union of signaling industry. In July 1996 SRS documents were published that formed the first baseline for the clinical specifications UNISIG provided for corrections and enhancement of the baseline specifications in April 1999. These baseline specifications has been tested by 6 railways since 1999 as part of the ERTMS.

## 5.4.2 Introduction

- The ETCs aims to remedy the lack of standardization in the area of signaling and train control systems, which contributes one of the major obstacles to the development of International Rail Traffic (IRT). Unifying the multiple signaling systems in use will bring increased competitiveness, better interworking of fright and passengers rail services, stimulate the European rail equipment market, reduce costs and improve the overall quality of the rail transport.
- ETCS is, infact, an autonomous train protection system based on cab signaling and spot and/or continuous track to train data transmission. It ensures trains operate safety at all time in providing safe movement authority directly to the driver through the cab display and in continuously monitoring the driver's actions.
- ETCS continuously calculates a safe maximum speed for each train, with can signaling for the driver and on-board systems that take control of the permissible speed is exceeded. For ETCS track side equipment and train borne systems need to be standardized according to the different ETCS levels.
- Thales offers a complete portfolio of ETCS solutions for turn key projects in the same way as for green field or brown field installations. Not only the obvious areas, such as track sides equipment or train borne installations are covered.
- At the same time, we will take case of other horizontal areas such as cyber security or special functionalities besides international specifications. For example, DAS, ATO IP based ratio communication and advanced diagnosis.

## 5.4.3 Definition

- The acronym ETCS stands for **European Train Control System**. **ETCS is the signaling and control component of the European Rail Traffic Management System (ERTMS).**
- It is a replacement legacy train protection system and designed to replace the many incompatible safety systems currently used by European Railways.
- **ETCs is the European Train Control System promoted by the European commission for use throughout Europe and specified for compliances with the High Speed and Conventional Inter Operability Directives (HSCOD).**
- **ETCs is the core signaling and train control component of ERTMS, the European Rail Traffic Management System.**

### 5.4.4 Levels

- ETCs is specified at four numbered levels as under :
  1. Level 0
  2. Level NT
  3. Level 1
  4. Level 2
  5. Level 3.

**1. Level 0 :**

- Level 0 applies, when an ETCs 3 fitted vehicle is used on a non-ETCS route. The train borne equipment monitors the maximum speed of that type of train. The train driver observes the track side signals.
- Since, signals can have different meanings on different railways, this level places additional requirements on driver's training. If the train has left a higher level ETCS, it might be limited in speed globally by the last **balises** encountered.

**2. Level 1 :**

- Level 1 is a cab signaling system that can be super-imposed on the existing signaling system, leaving the fixed signaling system (national signaling and track-release system) in place. Eurobalise radio Beacons pick up signal adapters and telegram coders (Line-side Electronics Unit - LED) and transmit them to the vehicle as a **movement authority** together with route data at fixed points.
- The on-board computer continuously monitors and calculates the maximum speed and the *braking curve* from these data. Because of the spot transmission of the spot data, the train must travel over the Eurobalise Beacon to obtain the next *movement authority.* In order for a stopped train to be able to move, when the train is not stopped exactly over a balise, there are optical signals that show permission to proceed. With the installation of additional Eurobalises (in fill balises) or Euroloop between the distant signal and main signal, the new proceed aspect is transmitted continuously.
- The Eoroloop is an extension of that Eurobalise over a particular distance that basically allows data to be transmitted continuously to the vehicle over cables emitting electromagnetic waves. A radio version of the Euroloop is also possible.

**3. Level 2 :**

- Level 2 is a digital radio-based system. *Movement authority* and other signal aspects are displayed in the cab for the *driver.* Apart from few indicator panel, it is therefore possible to *dispense* with the trackside signaling. However, the train detection and the train integrity supervision still remain in place at the trackside. Train movements are monitored continually by the radio block centre using this trackside-derived information.
- The *movement authority* is the transmitted to the vehicle continuously via GSM-R or GPRS together with speed information and route data. The Eurobalises are used at this level as passive positioning beacons or *electronic milestones.*
- Between two positioning Beacons, the train determines its position via sensors, axle transducers, accelerometer and radar. The positioning Beacons are used in this case as reference points for correcting distance measurement errors. The on-board computer continuously monitors the transferred data and the maximum permissible speed.

**4. Level 3 :**

- With level 3, ETCS uses beyond pure train protection functionality with the implementation of full radio-based *train spacing*. Fixed train detection devices (GFM) are no longer required. As with level 2, trains fixed their position themselves by means of positioning Beacons and via sensors (axle transducers, accelerometer and radar) and must be capable of determining train integrity on-board at the very highest degree of reliability.
- By transmitting the positional signal to the radio block centre, it is always possible to determine that point on the route the train has safely cleared. The following train can already be granted another *movement authority* upto this point. The route is thus no longer cleared in fixed track sections. In this respect, level 3 departs from classic operation with fixed intervals; given sufficiently short positioning intervals, continuous line-clear authorization is achieved and train head-ways come close to the principle of operation with absolute *braking distance* spacing (*moving block*).

- Level 3 uses radio to pass movement authorities to the train. Also it uses train reported position and integrity to determine, if it is safe to issue the *movement authority*. Solutions for reliable train integrity supervision are highly complex and are hardly suitable for transfer to offer models of fright rolling block. The Confirmed Safe Rear End (CSRE) is the point in rear of the point at the farthest extend of the safety margin.
- If the safety margin zero, the CSRE aligns with the confirmed Rear End. Some kind of **end-of-train device** is needed or special lines for rolling stock with included integrity checks like commuter multiple units or high-speed passenger units. Aghast train is a vehicle in the level 3 area that are not known to the Level 3 track side.

### 5.4.5 Train-Borne Equipments

- All the trains compliant with ETCS will be fitted with **on-board systems** certified by **Notified Bodies.** This equipment consists of wireless communication, rail path sensing, central logic unit, can displays and control devices for driver action.

  1. Man-machine interface.
  2. Specific transmission module.
  3. Balise transmission module.
  4. Odometric sensors.
  5. European vital computer.
  6. Error radio.
  7. Judicial recording unit.
  8. Train traffic unit.

### 5.4.6 Line-side Equipments

- **Line-side equipment is the fixed installed part of ETCS installation.** According to ETCS levels, the rail related part of installation is decreasing. While in Level 1 sequences with two or more if eurobalises are used as milestone application only.
- It is placed in Level 2 by mobile communication and more sophisticated software. In Level 3 even less fixed installation is used. In 2017, first possible tests for satellite positioning were there.

  1. Eurobalise
  2. Euroloop
  3. Lineside electronic unit
  4. Radio block centre

### 5.4.7 ERTMS

- The acronym ERTMS stands for **European Rail Traffic Management System**. **ERTMS is a single European signaling and speed control system that ensures inter-operability of the natural railway systems, reducing the purchasing and maintenance costs of the signaling systems as well as increasing the speed of trains, the capacity of infrastructure and the level of safety in rail transport.**
- ERTMS comprises of the ETCS, i.e., a cab signaling system that incorporates automatic train protection, the Global System for Mobile Communications for Railways (GSM-R) and operating rules. Technical specifications for ETCS and GSM-R are published in the Control Command and Signaling (CCS) Technical Specifications for Interoperability (TSI).
- GSM-R provides voice communications for train drivers and signalers, and provides data communication for ETCS. ERTMS and GSM-R rules are published in the Operation and Traffic Management (OTI).
- ERA plays the role of system design authority for ERTMS in that respect, it must establish a transparent process to manage with the contribution of the sector's representatives, any system changes.

## 5.5 DIGITAL CELLULAR SYSTEM STANDARDS

### 5.5.1 Brief History

- Analog cellular services in the United States were developed in the 1970 by AT & T and were widely used deployed by the late 1980s. They were implemented in a standard format so that all telephones worked on all analog cellular networks in the United States. However, analog cellular services become so popular that capacity was not adequate for future growth, particularly in metropolitan areas.

- As a result, digital cellular was developed to add capacity and advanced features. It offers features such as caller ID, call forwarding and three-way calling. In addition, many of the handsets have paging, long battery lives, and short messaging services integrated via **Liquid Crystal Display** (LCD).
- The advent of digital cellular and **Personal Communication Service** (PCS) led to price decrease and affordable cellular for residential consumers. In 1998, for the first time, the sale of digital handsets exceeded that of analog handsets. The digital cellular operates in the frequency range of 824 to 893 MHz.
- Digital cellular services are being implemented differently in the United States than in Europe. In 1987, the European Union choose a standard called global system for mobile communication (GSM) for delivering new digital wireless telephony. In the United States, the **Telecommunication Industry Association** (TIA) settled on a similar standard using TDMA. However, shortly many of the Ball telephone companies decided to use a newer technique of multiplexing, CDMA, which has greater capacity than TDMA. CDMA has greater capacity than that of TDMA.
- Thus, the United States started down the road with two different standards, both different than European's standards. In addition, the Nextel service works on a different frequency than that of PCs D-AMPS. Interestingly there were 7 different, incompatible analog types of cellular service in Europe before digital GSM was installed. Incompatibility is a problem, because customers with digital telephones cannot use their telephones, when they travel to places with incompatible cellular service.
- The manor challenge for digital cellular is that no one carrier has a complete nationwide network. PCs providers such as AT & T wireless and sprit PCS are still building out their networks, as is version wireless. Regional Ball Operating Companies, such as SBG and Bellsouth are merging to create networks throughout the country. Most importantly, there are pockets every where, particularly outside metropolitan areas, with poor service and poles in coverage.

## 5.6 DIGITAL AMPS (D-AMPS)

### 5.6.1 Brief History

- The evolution of mobile communication began in three different geographic regions namely, North America, Europe and Japan. The standards used in there regions wire quite independent of each other. The earliest mobile or wireless technologies implemented were wholly analogue, and the collectively known as *first generation* (1G) technologies.
- In Japan, 1G standards were *Nippon Telegraph and Telephone* (NTT) and the high capacity version of it (Hicap). The easy system used throughout Europe were not compatible to each other, meaning the later idea of a common *"European Union"* technological standard was absent at this time.
- The varians 1G standards in use in Europe included C-Netz in Germany and Austria, Comvia in Sweden, etc. North American Standard were AMPS and Narrow band AMPS (N-AMPS). Out of the 1G standards, the most successful was the AMPS system. AMPS developed by Bell laboratories in the 1970s and first used commercially in the United States in 1983 operates in the 800 MHz frequency band in the United States and is the most widely distributed analog standard.
- The success of AMPS kick-started the mobile age in North America. AMPS used FDMA which enabled each cell site to transmit on different frequencies allowing many cell sites to built near each other. AMPS also had many disadvantages, what triggered the search for a more capable system. The quest resulted in IS-54 on the first American 2G standards.

### 5.6.2 Types of D-AMPS

1. IS-54
2. IS-136

- IS-54 and IS-136 are second generation (2G) mobile phone systems known as D-AMPS and is further development of the North American 1G mobile system AMPS.

## 5.7 COMPARISON OF ANALOG AND DIGITAL CELLULAR SYSTEMS

**Table 5.1**

| Sr. No. | Parameter | Analog Cellular System | Digital Cellular System |
|---|---|---|---|
| 1. | Traffic channel | Voice use FM | Voice encoded in digital format. |
| 2. | Processing | More difficult | Easier by using modem. |
| 3. | Encryption | No security | Yes, security. |
| 4. | Noise | More noisy | Less noisy. |
| 5. | Error detection and correction | No such facility | Yes, such facility. |
| 6. | Voice clarity | Voice not clear | Clear voice. |
| 7. | Channel access | One channel to only one user | One channel shared by number of users. |
| 8. | Compatibility | Not compatible with other devices | Compatible with computers. |

## 5.8 IS-54 STANDARD

### 5.8.1 Description

- IS-54 is mostly referred to as TDMA, which is used in most standards. It is he first mobile communication system, which had provision for security and the first to employ TDMA technology. It is the first American 2G standard. In March 1990, the North American cellular network incorporated the IS-54B standards the first *North American dual mode digital cellular standard.*
- IS-54 standard increased the capacity, by cutting down voice channels from 30 kHz to 10 kHz. The other hands, it increased the capacity by digital, means using TDMA protocols. This technique separates calls by time, placing parts of individual conversations on the same frequency, one after the next. TDMA tripled call capacity.
- Using IS-54, a cellular carrier could convert any of its system's *analog* voice channels to *digital.* A dual mode phone user digital channels where available, and defaults to regular AMPS, where they are not. It was *backward compatible* with analogue cellular and indeed co-excited on the same some radio channels as AMPS. No analogue customers were left behind, they simply could not access IS-54 is now features IS-54 is supported *authentication*, a help in preventing fraud.

### 5.8.2 Definition

- **IS-54 is an extention of the analog AMPS and is commonly known as Digital AMPS (D-AMPS) and United States Digital Cellular (USDC).** The system uses hybrid FDMA and TDMA concept as it accepts 3 users per carrier.
- **IS-54 is the standard for the digital version of the United States AMPS (US AMPS) system.**
- **IS-54 is a second generation (2G) mobile phone systems, known as Digital AMPS (D-AMPS) and a further development of the North American 1G mobile system AMPS.**
- IS-94 was once prevalent throughout the Americas, particularly in the United States and Canada since the first commercial network was deployed in 1993.

### 5.8.3 Technology Specifications

- IS-54 employs the same 30 kHz channel spacing and frequency bands (824 to 849 MHz and 869 to 894 MHz) as AMPS. Capacity was increased over the preceding analog design by dividing each 30 kHz channel pair into 3 time-slots and digitally compressing the voice data, yielding 3 times the call capacity in a single cell.

- A digital system also made calls more secure because analog scanners could not access digital signals. IS-94 specifies 84 control channels, 42 of which are shared with AMPS. To maintain compatibility with the existing AMPS cellular telephone system, the primary forward and reverse control channels in IS-54 cellular systems use the same signaling techniques and modulation scheme (binary RSK) as AMPS.
- An AMPS/IS-54 infrastructure can support use of either analog D-AMPS phones. The access technique used for IS-54 is TDMA, which was the first U.S. digital standard to be developed. It was adapted by the TIA in 1992. TDMA subdivides each of the 30 kHz AMPS channels into 3 full-rate TDMA, which is capable of supporting a single voice call.
- Thus, TDMA could provide 3 to 6 times the capacity of AMPS traffic channels. TDMA was initially defined by the IS-54 standard and is now specified in the IS-13x series of specifications of the BIA/IIA. The channel transmission bit rate for digitally modulating the carrier is 48.6 kbps. Each frame has 6 time slots of 6.67 ms duration.
- Each time-slot carries 324 bits of information of which 260 bits are for the 13 kbps full-rate traffic data. The other 64 bits are overhead, 28 of these are for synchronization and they contain a specific bit sequence known by all receivers to establish frame alignment. Also, as with GSM, the known sequence acts as a training pattern to initialize an adaptive equalizer.
- The IS-54 system has different synchronization sequences for each of the 6 time-slots making up the frame, thereby allowing each receiver to synchronize to its own preassigned time-slots. An additional 12 bits in every time slot are for the SACCH (System Control Information). The **Digital Verification Colour Code** (DVCC) is the equivalent of the supervisory audio time used in the AMPs.
- There are 256 different 8-bit colour codes, which are protected by (12, 8, 3) *Hamming code*. Each base station has its own preassigned colour code, so any incoming interfering signals from distant cells can be ignored. The modulation scheme for IS-54 is 7C/4 Differential Quaternary Phase Shift Keying (DQPSK), otherwise known as differential 7t/4 4-PSK, or $\pi/4$ DQPSK.
- This technique allows a bit rate of 48.6 kbps with 30 kHz channel spacing to give a bandwidth efficiency of 1.62 bps/Hz.
- This value is 20% better than GSM. The major disadvantage with this type of linear modulation method is the power inefficiency, which translates into a heavier handheld portable and seven more inconvenient a shorter time between battery operation.

## 5.8.4 Call Processing

- A conversation's data bits makes up the data field. Six slots make up a complete IS-54 frame. Data in slots 1 and 4, 2 and 5, and 3 and 6 make up a voice circuit. DVCC for a unique 8-bit code value assigned to each cell. Each time slot in every frame must be synchronized against all others and a master clock for every think to word. Time slots for the mobile-to-base direction and constructed differently from the base-to-mobile direction. They essentially carry the same information, but are arranged differently.
- Notice that the mobile-to-base direction has a 6-bit ramp time to enable its transmitter time to get upto full power, and a 6-bit guard band during which nothing is transmitted. These 12 extra bits in the base-to-mobile direction are reversed for future use.
- Once a call comes in the mobile switches to a different pair of frequencies, a voice call radio channel, which the system carrier has made analog or digital. This pair carries the call, if an IS-54 signal is detected, it gets assigned a digital traffic channel, if one is available.
- The fast associated channel or FACCH performs handoffs during the call, with no need for the mobile to go back to the control channel. In case of high noise FACCH, embedded within the digital traffic channel over-rides the voice payload, degrading speech quality to convey control information.
- The purpose is the maintain connectivity. The slow associated control channel or SACCH does not perform handoffs, but conveys things like signal strength information to the base station. This IS-54 speech coder uses the technique called vector sum excited linear prediction (VSELP) coding.

- VSELP is a special type of speech coder, within a large class known as code extended linear prediction (CELP) code. The speech coding rate of 7.95 kbps achieves a reconstructed speech quality similar to that of the analog ones system using frequency modulation.
- The 1.95 kbps signal is then passed through a channel coder that loads the bit rate upto 13 kbps. The new half-rate coding standard reduces the overall bit rate for each call to a 6.5 kbps, and should provide comparable quality to the 13 kbps rate. This half data gives a channel capacity 6 times that of analog AMPs.

## 5.9 IS-136

### 5.9.1 Introduction

- In earlier cellular communication systems the voice channel was defined only by a frequency. Thus, more calls can be transmitted by sharing one transmission frequency. Thus, more calls can be transmitted by sharing one transmission frequency. The dual mode system uses a **Digital Traffic Channel** (DTC) or an **Analog Voice Channel** (AVC).
- The dual mode system uses only an **Analog Control Channel** (ACC) to control transmission on the voice and traffic channels. IS-136 has the DTC, AVC, ACC, and adds a **Digital Control Channel** (DCCH). IS-136 expands the capability of IS-54B to include the following :
  1. Sleep mode for decreased battery using during non-talk times.
  2. Public, private and semi-private cells, such as pico cells in office buildings and personal base stations.
  3. **Short Message Service** (SMS) for both point-to-point and broadcast information.
  4. Creating improved security using DCCH and authentications.

### 5.9.2 Definition

- **IS-136 is a mobile communications Interium Standard, which extends the functions of the dual mode system standard IS-54B.** It is also known as NADC/TDMA (**North American Dual Mode cellular**, TDMA).
- NADC/TDMA is a dual mode, **Full Duplex Cellular Communications** (FDCC) system, in which each voice channel can be defined by both a frequency and a time-slot.

### 5.9.3 Basic Features

1. Time-slots per channel : 6.
2. Users per channel : 3 full rate, 6 half rate.
3. Modulation : Digital.
4. Analog : FM.
5. Data structure : TDMA.
6. Speech coding : VSELP.
7. Modulation data rate : 24.3 K symbols/s
8. Control channels : Digital and Analog control.
9. EIA/TI standards : IS-136 and 136.2 for symbols, IS-137 for mobile stations and IS-138 for base stations.

### 5.9.4 Radio Aspects

1. Channel bit rate : 48.6 kbps
2. Frame duration : 40 ms
3. Each slot 324 bits, 260 user data
4. Full rate and half rate : VSELP
5. Source rate : 7.95 kbps
6. DQPSK : $\pi/2$ shifted
7. Roll-off factor : 0.35
8. 1.62 bps.Hz
9. COPD pocket data transmission : Idle voice channels.

## 5.10 GLOBAL SYSTEM FOR MOBILE (GSM)

### 5.10.1 Brief History

- In 1983, work began to develop a European standard for digital cellular voice telecommunications, when the European Conference of Portal and Telecommunications Administrations (CEPT) setup the **Groups Special Mobile** (GSM) committee and later provided a permanent technical-support group based in Paris.

- Five years later, in 1987, 15 representatives from 13 European countries signed a *Memorandum of Understanding in Cepenhagen* to develop and deploy a common cellular telephone system across Europe and EU rules were passed to make GSM a mandatory standard. The decision to develop a continental standard eventually resulted in a unified, open, standard-based network, which was larger than that in the United States.
- In February 1987, Europe produced the first agreed GSM Technical Specification, Monitors from the 4 big EU countries cemented their political support for GSM with the Bonn Declaration on Global MoU was tabled for signature from across Europe to pledge to invest in new GSM networks to an ambitious common date.
- In this short 33 weeks period, the whole of Europe (countries and industries) had been brought behind GSM in a rare unity and speed guided by 4 public officials, Armin Siberhon (Germany), Stephen Temple (UK), Phillipe Duppis (France) and Renzo Faulli (Italy). In 1989, the group special Mobile Committee was transferred from CEPT to the **European Telecommunications Standards Institute** (ETSI).
- In parallel *France* and *Germany* signed a joint development agreement in 1984 and were joined by *Italy* and *UK* in 1986. In 1986, the European commission proposed reserving the 900 MHz spectrum band for GSM. The former *Finish* Prime Minister *Harry Helkeri* mode the world's first GSM call on July 1, 1991, calling *Koarnia Suanio* (deputy mayor of the city Temper) using a network built by **radio linja.**
- The following year saw the sensing of the first **Short Messaging Service** (SMS) or Text Message and *Vodafone*, UK and Telecon, Finland signed the first international *roaming* agreement. Work began in 1991 to expand the GSM standard to the 1800 MHz frequency band and use first 1800 MHz network became operational in the UK by 1993, called and DCS 1800.
- Also that year, Telecom Australian become the first network operator to deploy a GSM network outside Europe and the first practical band-held GSM mobile phone become available. In 1995, fax, data and SMS messaging services were launched commercially, the first 1900 MHz network become operational in the United States and GSM subscribers world wide exceeded 10 million. In the same year the GSM Association formed.
- Pre-paid GSM, SIM cards were launched in 1996 and world wide GSM subscribers passed 100 million in 1998. In 2000, the first commercial GPRS services were launched and the first GPS-compatible handsets became available for sale. In 2001, the first UMTS (W-CDMA) network was launched, a 3G technology that is not part of GSM.
- Worldwide GSM subscribers exceeded 500 million. In 2002, the first Multimedia Message Service (MMS) was introduced and the first GSM network in the 800 MHz frequency-band become operational. EBGE services first became operational in a network in 2033, and the number of world wide GSM subscriber's exceeded 1 billion in 2004.
- By 2005, GSM networks accounted for more than 75% of the world wide cellular network market, serving 1.5 billion subscribers. In 2005, the first HSDPA capable network also became operational. The first SUPA network launched in 2007, HSPA (High-Speed Packet Access) and its uplink and downlink versions are 3G technologies not part of GSM. World wide GSM subscribers exceeded 3 billions in 2008.
- The GSM association estimated in 2011 that technologies defined in the GSM standard served 80% of the mobile market, encompassing more than 5 billion people across more than 212 countries and territories, making GSM the most ubiquitous of the many standards by cellular networks.
- GSM is a second generation (2G) standard employing TDMA spectrum sharing, issued by the European Telecommunications Standard Institute (ESTI). The GSM standard does not include the 3G-UMTS (Universal Mobile Telecommunications System) code CDMA technology not the 4 GLTE orthogonal frequency FDMA (OFDMA) technology standards issued by the 3 GPP.
- GSM, for the first time, set a common standard for Europe for wireless network. It was also adopted by many countries outside Europe. This allowed the subscribers to use other GSM networks that have roaming agreement with each other. The common standard reduced research and development (R&D) costs, since hardware and software could be sold with only minor adaptations for the local market.

- **Telstra** in Australia shut down its 2G GSM network on December 1, 2016, the first mobile network operator to decommission a GSM network. The second mobile provider to shut down its GSM network on January 1, 2017 was AT&T **mobility** from the United States. **Optus in Australia** completed the shut down its network covering Western Australia and Northern Territory had earlier in the year been shut down in April 2017. Singapore shut down 2G services entirely in April 2017.

## 5.10.2 Introduction

- GSM communication is a wireless communications system that uses digital cellular radio, communication to provide voice, data, and multimedia application's communication services. It covers wide area of range of mobility.
- A GSM system coordinates the communication between mobile telephone units (mobile stations), base stations, system (cell sites), and switching systems (MSC or MTSO). The bandwidth of each GSM radio channel is 200 kHz wide and which are further divided into frames that hold and time slots. GSM was originally named as Groupes Speciate Mobile.
- The GSM system includes mobile telephones (mobile stations), radio towers, (base stations system) and interconnecting switching systems (MTSO or MSC).
- The GSM system allows upto 8 to 16 voice users to share each channel and there may be several radio channels per radio transmission site (cell site) in every honey comb should be hexagonal cell.
- GSM was the world's first cellular system to specify digital modulation and network level architectures and services.
- The Conference of European Postal and Telecommunications Administrations (CEPTA), the European group began to develop the GSM standards for TDMA system in June 1982.
- GSM was originally developed to serve two objectives :

1. Pan - European roaming, which offers compatibility throughout the European continent.
2. Interaction with the Integrated Service Digital Network (ISDN), which offers the capability to extend the single subscriber line system to a multi service system with various services which are currently offered only through diverse communications network.

- GSM was first introduced into the European market in the year 1991. The first commercial GSM system, called $D_2$, was implemented in Germany in the year 1992.
- Before the GSM system was available, most countries throughout the world used cellular mobile communication systems that were often incompatible with each other. Most cellular mobile telephone units could only operate on a single type of cellular system, so most customers could not roam to neighbouring countries.

## 5.10.3 Definition

- **The Global System for Mobile (GSM) communications is a standard developed by the European Telecommunications Standard developed by the European Telecommunications Standard Institute (ETSI) to describe the protocols for second generation (2G) digital cellular networks used by mobile devices such as mobile phones and tablets.**
- The Global System for Mobile (GSM) communication is one of the most popular mobile communication standards. The GSM communication uses cellular networks.
- **GSM is a second generation (2G) cellular (mobile radio) system standard that was developed to solve fragmentation problems of the first cellular systems.**
- GSM is a globally accepted standard for digital cellular communication. GSM is the name of a standardization group established in 1982 to create a common European mobile standard that would formulate specifications for a European mobile cellular radio system operating at 9030 MHz.

### 5.10.4 Evolution of GSM Standards

- The basic evolution of the GSM industry standards is as follows:
- In 1990, phase 1 of the GSM specifications was completed, including basic voice and data communication cellular services. At that time, work began to adopt the GSM cellular communication specification to provide service in the 1800 MHz frequency band range. This 1800 MHz standard known as DCS 1800 is used for the Personal Communication Network (PCN).

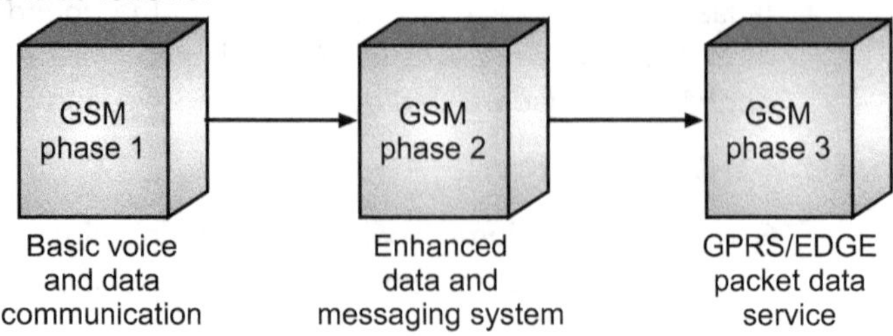


**Fig. 5.2 : Evolution of GSM standards**

- Phase 2 of the GSM and DCS 1800 specifications which added advanced :
  (a) Short messaging.
  (b) Microcell support services, and
  (c) Enhanced data transfer capability are now complete.
- Phase 2 added the advanced information services and pocket data transmission capability.
- The GSM association helps in (a) promotion, (b) protection, and (c) evolution of GSM technology and products throughout the world.
- Information about the GSM association is available at www.gsmworld.com.
- GSM association members include : (a) Mobile operation, (b) Manufacturers, and (c) Suppliers.
- The GSM standardization progress is as under :
  (a) Originally the GSM development group was hosted by CEPT.
  (b) GSM technology basics were created in 1987 and then in 1989.
  (c) ETSI becomes the managing body.
  (d) In 1990, the first GSM specification was released (more than 6000 pages of specifications).
- The third generation (3G) partnership project (3GPP) group was formed in 1998, to create the next evolution of mobile specification in cellular mobile communication.
- The 3GPP has now taken over the management of GSM specifications. GSM specifications and (evolved versions of the specifications) is available at www.3GPP.org.

### 5.10.5 Features

1. Improved spectrum efficiency
2. International roaming
3. Compatibility with ISDN
4. Support for new services
5. SIM phonebook management
6. Fixed dialing number
7. Real time clock with alarm management
8. High-quality speed
9. On-air privacy for users

### 5.10.6 Characteristics

- The important characteristics of GSM standards are as under :
  1. Fully digital system utilizing the 900 MHz frequency band.
  2. TDMA over radio carriers (200 kHz carrier spacing).
  3. 8 full rate of 16 half-rate TDMA channels per carrier.
  4. User/terminal authentication for fraud control.

5. Encryption of speech and data transmission over the radio path.
6. Full international roaming capacity.
7. Low speed data services (upto 9.6 kb/s).
8. Compatibility with ISDN for supplementary services.
9. Support of Short Message Service (SMS).

## 5.10.7 Specifications

- The GSM system has several specifications as under :
  1. Roaming in European countries.
  2. Connection to ISDN through RA box.
  3. Use of Subscriber Identity Module (SIM) card.
  4. Control of transmission power.
  5. Frequency hopping.
  6. Discontinuous transmission.
  7. Mobile assisted handover.
  8. On-the-air privacy for the user.

## 5.10.8 GSM Services

- GSM provides integrated services for voice and data. There are three types of services delivered by a GSM system as : (1) Teleservices, (2) Bearer (or data) services and (3) Supplementary ISDN services.
- They also include subservices. The various services offered by a GSM system are as under :

**1. Teleservices or Telephone Services :**

(a) Regular phone calls
(b) DTMF
(c) Emergency call
(d) Fax
(e) SMS and MMS
(f) Call broadcast
(g) Voice mail
(h) Fax mail
(i) Video text and Tele text

**2. Bearer (or data) Services :**

(a) Connections to PSTN
(b) Connections to ISDN
(c) Connections to PDN
(d) Full duplex
(e) Data transmission
(f) Package data

**3. Supplementary (ISDN) Services :**

(a) Call forwarding,
(b) Caller line ID,
(c) Connected line ID,
(d) Call barring,
(e) Call waiting,
(f) Call hold,
(g) Connected line ID,
(h) Closed user group,
(i) Multy party (teleconferencing),
(j) Call charge advice,
(k) Calling line presentation restriction.

## 5.10.9 Description of GSM Services

- The available services in GSM cellular systems are as explained below :

**1. Teleservices :**

**(a) Regular phone calls :** Common GSM application.

**(b) Dual Tone Multi Frequency (DTMF) :** This is the signaling method to allow the user to control a device connected to phone line using the keypad of the mobile station for handset.

**(c) Emergency call :** If cell is full, then emergency call leads to interruption of another connection.

**(d) Fax :** Sending fax with regular GSM connection. To send and receive the Fax via GSM, Terminal Adaptor Function (TAF) provide general interface between GSM devices and other devices.

(e) **SMS :** Short Message Service (SMS) consists of upto 160 alphanumerical characters.

(f) **Cell broadcast :** A cell broadcast transmits news regarding local traffic situation or any social violence in that area.

(g) **Voice mail :** It allows the user/customer to forward incoming calls to a service center of GSM network that acts like an answering machine. After listening to a message from the subscriber the calling party may leave a voice message which the subscriber can retrieve by connecting this service center.

(h) **Fax mail :** Service allows the subscriber to forward incoming faxes to special center in the GSM network. Then faxes can be retrieved by the customer by connecting to the center with a PSTN connection or GSM connection.

2. **Bearer (Data) Services :**

(a) **Connections to the PSTN :** Subscriber can connect to modems connected to an analog PSTN landline.

(b) **Connections to the ISDN :** Digital information can be connected to digital network like Integrated Services Digital Network (ISDN).

(c) **Connection to the packet switched networks :** Subscriber can also access packed switched network like PDN (Packet Data Network).

3. **Supplementary Services :**

(a) **Call forwarding :** Subscriber can select under which conditions call to higher mobile numbers are forwarded to another mobile number like :

(i) always

(ii) in case the MS cannot be reached

(iii) subscriber is making or receiving another call

(iv) subscriber does not answer after a specified number of rings to that number.

(b) **Blocking of incoming calls :** When user has to pay part of the charges for incoming calls (roaming out of the original network).

(c) **Advice of charges :** Subscriber may be able to access an estimation of the call charges.

(d) **Call hold :** The subscriber may put a connection on hold to make or receive another call and then continue with first call connection.

(e) **Call waiting :** During a call, the customer may be informed about another incoming call. User may either answer this call by putting the other call on hold or reject it. This type of feature is available to all circuit switched connections to accept emergency calls.

(f) **Conference calls :** The feature enables connection to multiple customers/users simultaneously. It is only possible for normal speech communication.

(g) **Caller ID :** The incoming phone number is displayed.

(h) **Closed groups :** Customers/users in GSM ISDN and other networks may be defined as a specified user groups. Members of this group can be allowed to make only within groups.

## 5.11 GSM SYSTEM ARCHITECTURE

### 5.11.1 Architectural Diagram

- Fig. 5.3 shows the architecture of a GSM system. It consists of three major interconnected subsystems that interact between them themselves and with the users through certain network interfaces. The subsystems used in the GSM architecture are the Base Station Subsystem (BSS) or Radio Subsystem (RSS), Network and Switching Subsystem (NSS) and Operation Subsystem (OSS).
- The GSM is a combination of FDMA and TDMA.

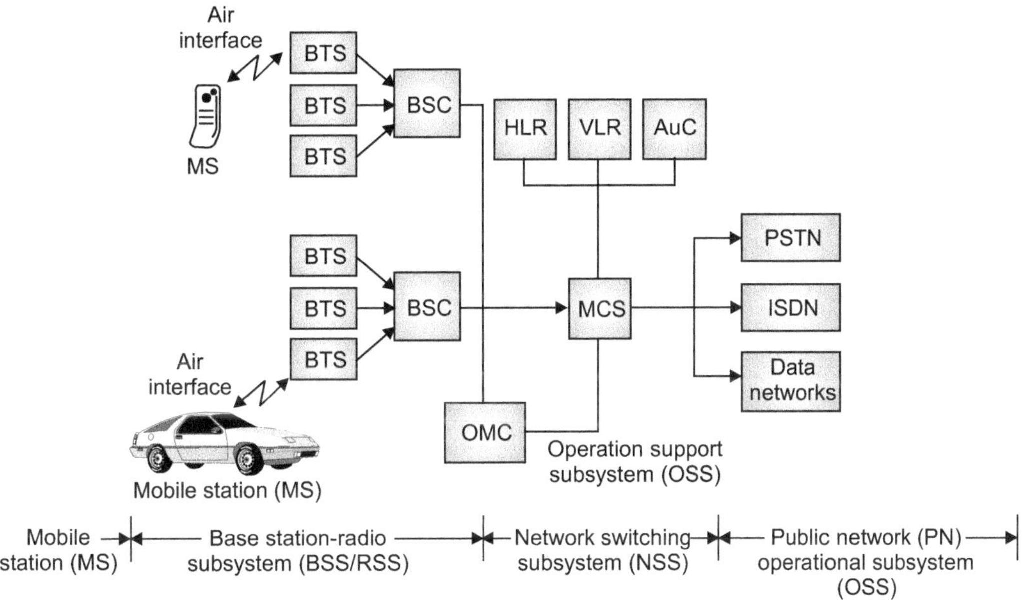


**Fig. 5.3 : GSM system architecture**

## 5.11.2 Description of Subsystems

- Consider the GSM system architecture as shown in Fig. 5.3. It consists of four major subsystem blocks as under :
  1. Mobile Station (MS).
  2. Base Station Subsystem (BSS).
  3. Network and Switching Subsystem (NSS).
  4. Operating Support Subsystem (OSS).
- The functions of each subsystem block are discussed as under :
  1. **Mobile Station (MS) :** The mobile station (MS) is also a subsystem, but it is usually considered to be a part of the Base Station Subsystem (BSS) for architecture purposes. It communicates with the BSS over the radio air interface. It is used to support the connections of the external terminals such as a PC or Fax.
  2. **Base Station Subsystem (BSS) :** The base station subsystem (BSS) is also known as *Radio Station Subsystem* (RSS). It consists of number of base station controllers (BSCs) and Base Transreceiver Stations (BTSs). Each BSC is connected to a number of BTSs, which in turn provides radio interface for devices. The BSS gets connected to MS through a radio (i.e. air) interface and it also gets connected to the network and switching subsystem (NSS). The GSM uses the open system interconnection (OSI) model.
  3. **Network and Switching Subsystem (NSS) :** The NSS consists of a number of Mobile Service Switching Centres (MSCs). Each MSC consists of the NSS interfaced to a number of BSCs in the BSS. There are also Home Location Resistors (HLRs), and Visitor Location Registers (VLRs). The NSS uses the Intelligent Network (IN) and the signaling is one of the main switching functions of GSM. The NSS is designed to manage the communication between GSM and other communication users.
  4. **Operation Support Subsystem (OSS) :** The OSS consists of the Authentication Centre (AuC), Equipment Identity Register (EIR) and Operation and Maintenance Centre (OMC). The areas coming under OSS are as under :
     1. Network operation and maintenance.
     2. Charging and billing.
     3. Management of mobile equipment.

### 5.11.3 Functions of Each Subsystem

- The functions of various subsystems of GSM system architecture are discussed as under :
  1. **Mobile Station (MS) :** It communicates with the BSS over the radio air interface. It supports the connections of the external terminals such as PC or Fax.
  2. **Base Station Subsystem (BSS) :** It is responsible for all functions related to the radio channel management. It performs the following functions :
     (i) Management to radio channel configuration with respect to use a speech.
     (ii) Management to data or signaling channels.
     (iii) Management to allocation and release of channels for call set up.
     (iv) Management to release control frequency hopping and transmit power at the mobile station.
  3. **Network and Switching Subsystem (NSS) :** It manages the communication between GSM and other communication users. The signaling is main function for carrying information between the networks, i.e. the base station and the mobile station. The signaling is related to the operating parameters such as location area identity, cell identity, call set up (e.g. paging), location updating etc.
  4. **Operation Support Subsystem (OSS) :** It operates and maintains the network. It keeps the records of charging and billing. It manages the various mobile equipments.
  5. **Home Location Register (HLR) :** Each subscriber assigned IMSI to identify home user. Database contains IMSI (International Mobile Subscriber Identity), prepaid-postpaid, roaming restrictions, supplementary services. Database contains subscriber and location information. Permanent database about mobile subscribers in a large service area (generally one per GSM network operator).
  6. **Visitor Location Register (VLR) :** It controls the mobiles roaming in its area. It updates whenever new MS enters its area, by HLR database. Temporary database which stores IMSI and customer information for each roaming subscriber visiting the coverage area of particular MSC.

## 5.12 GSM SYSTEM BLOCK DIAGRAM

### 5.12.1 Block Diagram

A block diagram of a typical GSM system is shown in Fig. 5.4.

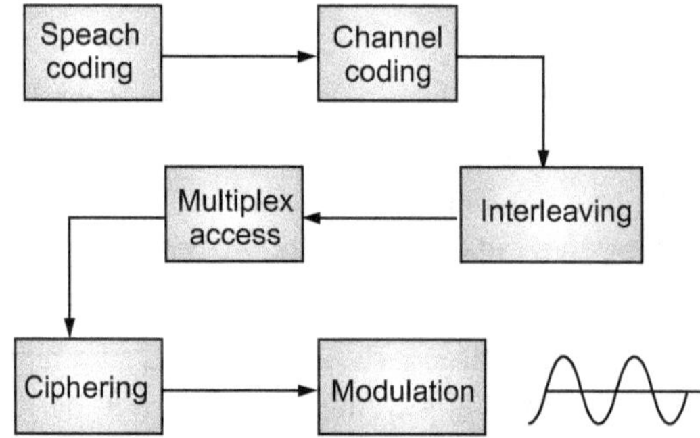


**Fig. 5.4 : Typical GSM system block diagram**

### 5.12.2 Description of Each Block

- To transform the speech that enters the handset into radio waves that are transmitted to the base station.
- GSM performs the following operation. To decode the message all of the above stages must be undergone. These stages are shown in Fig. 5.4 for clarity.
  **(1) Speech coding :** GSM is a digital communication standard but voice is analog and therefore it must be first converted to a digital bit stream. GSM uses Pulse Coded Modulation (64 kbps) to digitize voice and then uses the Full Rate Speech Code to remove the redundancy in the signal and achieve a bit rate of 13 kbps.

(2) **Channel coding :** Once the voice signal has been coded into a digital bit stream, extra bits are added to the bit stream that the receiver can recognize and correct errors in the bit stream which could have occurred during the transmission. GSM uses a technique called convolutional coding.

(3) **Interleaving :** Interleaving is the process of rearranging the bits. Interleaving allows the error correction algorithms to correct more number of errors that could have occurred during transmission by interleaving the code. There is less possibility that a whole chuck of code is lost.

(4) **Multiple access :** GSM allows many users to use their cell phones at the same time. GSM uses a combination of Time Division Multiplex Access (TDMA) and Frequency Division Multiplex Access (FDMA) to share the limited bandwidth that is provided by regulators to the service providers. FDMA divides the spectrum with small slices and then each frequency slice is separated in time into many blocks by TDMA. The transmission of the voice signal is no longer continuous because of the division of the frequency slice in time, but the data is transmitted in burts. The burst assembly operation take the final encoded data and groups it into burts.

(5) **Ciphering :** Ciphering is used to encrypt the data so that no one can hear the conversation of another user. In GSM, the two parties in encrypting and decrypting the data are the Authentication Center (AuC) and the SIM card in the mobile phone. Each SIM card holds a unique secret key which is known by the AuC. The SIM card and AuC then follow a couple of algorithms to first authenticate the user and then encrypt the data. For authentication, the AuC sends a 128 bits random number to the mobile phone. The SIM card uses its secret key algorithm to perform a function on the random number and sends back the 32-bit result. Since the AuC knows the SIM cards special key, it performs the same function and checks that the results obtained from the mobile phone match the result it obtained. If it does, the mobile user is authenticated. Once authentication has been performed, the random number and the secret key are used in the A8 algorithm to obtain a 64-bit ciphering key. This ciphering key is used with the TDMA from number in the A8 to generate a 114-bit sequence. Note that ciphering key is constant throughout a conversation but the 114-bit sequence is different for every TDMA frame. The 114-bit sequence is XORed with the two 57-bit blocks in a TDMA burst. The only user that can decrypt the data is the mobile phone or the AuC. Since they are the only ones that have access to the secret key, which is needed to generate the ciphering key and the 114-bit sequence. Note that the A3, A5 and A8 algorithms are known to the public domain, however, some information about A5 has been leaked. It is known that A5 has a 40-bit key length. This allows for the encryption to be broken in a matter of days; but since, cellular cells have a short lifetime, the weakness of algorithm is not an issue.

(6) **Modulation :** The original analog voice signal has been digitized into interleave grouped and encoded and the digital data is ready to be transmitted. The digital bit stream must be encoded in a pulse and transmitted over radio frequencies. Modulation changes the '1' and '0's in a digit representation to another representation that is more suitable for transmission over airwaves.

## 5.13 GSM CHANNELS

### 5.13.1 Concept

- **In GSM cellular communication, logical channels are a portion of a physical channel that is used for a particular (logical) communication purpose.**
- The physical channel may be divided in time frequency or digital coding to provide for these logical channels.

### 5.13.2 Need

- In practice, a multimode terminal used by a third generation (3G) mobile communication system network will have to scan for a suitable frequency band/channel, identify the applicable radio and standard and select from among the set or available services.
- If it develops that a very large number of frequency bands need to be scanned and the many standards need to be searched, registering such a roaming multimode terminal by means of a systematic scanning procedure will become very inefficient, tending to degrade the quality of service from the user's perspective.

- This problem could be alleviated by using a common physical or logical broadcast channel called the global radio control channel.
- Therefore *there is a necessity of logical channel in GSM system to scan a single frequency or a small range of frequencies and thereby find the required information on available networks/standards and services.*

## 5.13.3 Radio Interface

- The **radio interface** is the general of the connection between the Mobile Station (MS) and the Base Transceiver Station (BTS). It utilizes the TDMA concept with one TDMA frame per carrier frequency.
- Each frame consists of 8 time-slots (TS). The direction from BTS to MS is defined as the **downlink** and opposite direction as the **uplink.**

## 5.13.4 Channel Types

1. Physical channels
2. Logical channels.

**1. Physical Channels :**

- When a Mobile Station (MS) and Base Transceiver Station (BTS) communicate, they do so on a specification of Radio Frequency (RF) carriers, one for the up-link and the other for the downlink transmissions, and within a given time slot. This combination of time slot and carrier frequency is called a **physical channel.**
- One RF channel will support 8 physical channels in time-slot through 7.

**2. Logical Channels :**

- The data, whether user traffic or signaling information are mapped onto the physical channels by defining a number of logical channels.
- A logical channel will carry information of a specific type and a number of these channels may be combined before being mapped onto the same physical channel e.g., speech.
- The channel types are again classified in two types :

  1. Traffic Channel (TCH) 2. Control Channels.

- The GSM global (logical) channel is required in third generation (3G) GSM system to facilitate multimode terminal operation and world wide roaming.
- There are two types of GSM logical channels as under :

1. GSM Traffic Channels (TCHs).
2. GSM Control Channels (CCHs).

Fig. 5.5 shows the various logical channels associated with D-AMPs when the DCC is used.

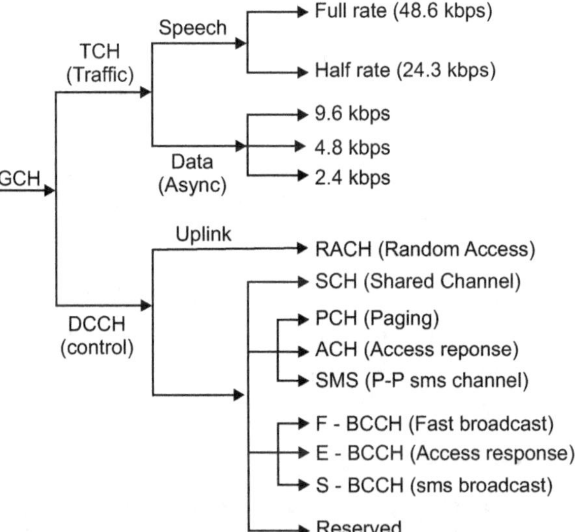


**Fig. 5.5 : Logical channel associated with D-AMPs**

## 5.13.5 GSM Traffic Channels

**1. Concept :**

- The traffic (data) channels carry digitally encoded user speech (or user data) and reverse link.
- The voice is coded using codec.

- The word codec is the short form for coder-decoder, which is a circuit that codes analog signals into digital signals and decodes digital signals into analog signal according to various coding and decoding algorithms.
- The error correction bits, Cyclic Redundancy Check (CRC), and redundant bits are then appended and data interleaving is performed.
- An application of data interleaving is as follows : a user talking on a mobile phone feels uneasy when there are abrupt periods of silence due to abrupt losses of connectivity.
- A technique to provide a relief is to interleave an induced background noise or roadside traffic noise.
- When a voice activity detector is not detecting any voice traffic from the other end due to the user not talking or some processing related delays or losses of connectivity, interleaving at the leastener's end helps the user, who does not feel uneasy due to abrupt intervals of silence.

**2. Types :**

- There are three types of voice (i.e. speech) traffic, namely voice at (i) full rate of 13 kbps, (ii) half data-rate speech, and (iii) enhanced full-rate speech.
- *The user data as subscriber data is also called traffic data.*

**(i) Full-Rate TCH :** The full-rate traffic channels (speech and data) are supported by the following channels :

**(a) Full-Rate Speech Channels (TCH/FS) :**

- It is a traffic channel/full-rate set for transmission of speech.
- The full-rate speech channel carries user speech (voice) which is digitized (i.e., coded) by a codec at a raw data rate of 13 kbps.
- With the addition of bits to the digitized voice after coding, the data rate is enhanced to 22.8 kbps.
- When transmitted as full-rate, the user data is contained within one Time Slot (TS) per frame.

**(b) Full-Rate Data Channel for 9600 bps (TCH/F9.6) :**

- The full-rate traffic data channel carries raw user data which is sent at 9600 bps.
- With additional forward error correction coding applied by the GSM standard, the 9600 bps data is sent at 22.8 kbps.

**(c) Full-Rate Data Channel for 4800 bps (TCH/F4.8) :**

- The full-rate traffic data channel carries raw user data which is sent at 4800 bps.
- With additional forward error correction coding applied by the GSM standard, the 4800 bps is sent at 22.8 kbps.

**(d) Full-Rate Data Channel for 2400 bps (TCH/F2.4) :**

- The full-rate traffic data channel carries raw user data which is sent at 2400 bps.
- With additional forward error correction coding applied by the GSM standard, the 2400 bps is sent at 22.8 kbps.

**(ii) Half-Rate TCH :** The half-rate speech (voice) and data channels are supported by the following :

**(a) Half-Rate Speech Channel (TCH/HS) :**

- It is a traffic channel/half rate set for transmission of speech.
- The half-rate speed channel is designed to carry speech, which is sampled at a rate half that of the full-rate channel.
- The voice is coded with a coder at 6.5 kbps and after the addition of error correction bits, the data rate is enhanced to 11.4 kbps and transmission takes place at half speed. The available data rate is 22.8 kbps.
- When transmitted as half-rate, the user is mapped onto the same Time Slot (TS), but is sent in alternate frames. That is, two half-rate channel users would share the same time slot, but would alternately transmit during every other frame.
- The advantage of this sort of transmission is that double voice signals can now be transmitted. However, this sort of voice data results in degradation of voice quality.

**(b) Half-Rate Data Channel for 4800 bps (TCH/4.8) :**

- The half-rate traffic channel carries raw user data which is sent at 4800 bps. With additional forward error correction coding applied by the GSM standard, the 4800 bps data is sent at 11.4 kbps.

**(c) Half-Rate Data Channel for 2400 bps (TCHs/H2.4) :**

- The half-rate traffic channel carries raw user data which is sent at 2400 bps.
- With additional forward error correction coding applied by the GSM standard, the 2400 bps data is sent at 11.4 kbps.

**(iii) Enhanced Full-Rate Channel (TCH/EFR) :**

- It is a traffic channel/enhanced full-rate set for transmission of speech and data.
- The voice is coded with another enhanced coding technique employing a code.
- The Expanded Full Rate (EFR) gives an enhanced voice quality but has limited error correction bits because the data rate is limited to 12.8 kbps.
- It is advantageous since the voice quality is upgraded in those cases where the transmission error rate is small.
- A codec may function in automatic mode and code the voice as TCH/FS or TCH/EFR depending on the transmission error rate detected in the bursts.

## 5.13.6 GSM Control Channels

**1. Types :**

- The 184-bit packet from data link layer is formatted for the data burst bits.
- The 184-bits are added after 40 party bits, four bits and 224 hold convolution coding bits to result in the 456-bit packet.
- There are three main types of control channels in the GSM system. They are :

  (i) Dedicated Control Channel (DCCH). (ii) Broadcast Control Channel (BCCH).

  (iii) Common Control Channel (CCCH).

**(i) Dedicated Control Channel (DCCH) :**

- Dedicated Control Channel (DCCH) GSM and like traffic channels, they are bidirectional and have the same format and functions on both the forward and reverse links.
- Like traffic channels, DCCH may exist in any Time Slot (TS) and on any Absolute Radio Frequency Channel Number (ARFCN) except TS 0 of the BCH ARFCN.
- There are three types of DCCH. They are : (i) Standalone Dedicated Control Channel (SDCCH), (ii) Slow-Associated Control Channel (SACCH) and (iii) Fast Associated Control Channel (FACCH).

**(a) Standalone Dedicated Control Channels (SDCCHs) :**

- A mobile station sends traffic channel only after a cell set up. A bidirectional communication channel is present between the Base Transceiver Station (BTS) and the mobile station before the TCH traffic starts. It is called the *standalone DCCH*, i.e., SADCCH.
- The SDCCH carries signaling data following the connection of mobile with the base station and just before a traffic channel assignment is issued by the base station.
- The SDCCH ensures that the mobile station and the base station remain connected while the base station and mobile station centre verify the subscriber unit and allocate resources for the mobile.
- The SDCCH can be thought of an intermediate and temporary channel which accepts a newly completed call from the Broad Band Channel (BCH) and hold the traffic, while waiting for the base station to allocate a traffic channel.
- The SDCCH is used for the registration, authentication and other requirements.
- It is used to send authentication and alert messages (but not voice) as the mobile synchronize itself with the frame structure and wait for a transfer channel.
- The SDCCHs may be assigned their own physical channel or may occupy $TS_0$ of the broadband channel, if there is low demand for broad band or common control channel traffic.

**(b) Slow Associated Control Channel (SACCH) :**

- The SACCH is always associated with a traffic channel area SDCCH and maps onto the same physical channel.
- Thus each ARFCN systematically carries SACCH data for all of its current users.
- As in the United States Digital Cellular (USDC) standard, the SACCH carries general information between the mobile station and base transceiver.
- On the forward link, the SACCH is used to send slow but regularly changing control information to the mobile, such as transmit power level instructions and specific timing advance instructions for each user on the ARFCN.
- The reverse SACCH carries information about the received signal strength and quality of the traffic channel, as well as broadcast channel measurement result from neighbouring cells.
- Total 782 bits are sent as dedicated control channel data in 1 second in case of slow associated standalone DCCH (SADCCH). 950 bps can be sent as a control data slot in a traffic mainframe.

**(c) Fast Associated Control Channels (FACCHs) :**

- When more than 782 bits are to be sent per second, then the traffic channel part of the data burst can be used.
- Then DCCH is called FACCH. It carries urgent messages, and contains essentially the same type of information as the SDCCH.
- A FACCH is assigned whenever a SDCCH has not been dedicated for a particular user and there is an urgent message such as a hand-off request.
- The FACCH gains access to a time slot by *stealing* frames from the traffic control channel to which it is assigned.
- This is done by setting two special bits, called stealing bits, in a traffic forward channel burst.
- If the stealing bits are set, the time slot is known to contain FACCH data, not a traffic channel, for that frame.
- Only FACCH and SACCH data is transmitted in a traffic multiframe. The traffic channels transmit only after the call set up and transmit the user/subscriber data through the base transceiver.
- A separate multiframe is required for the control channels, other than FACCH and SACCH. That mainframe is called *control mainframe.*

**(ii) Broadcast Control Channels (BCCHs) :**

- A Base Transceiver Station (BTS) needs to broadcast the frequency and cell identity. A Broadcast Control Channel (BCCH) is used for that.
- A BTS needs to broadcast the information regarding frequencies and sequence options for hopping that can be assigned to the mobile stations (MSs) in the cell to cell the MSs.
- This enables MS to get an avoidable radio-carrier frequency channel and transmit with different frequencies on different hops and synchronize with the BTS.
- The synchronization and frequency correction bursts also use the BCCH.
- The BCCH operates on the forward link of a specific ARFCN within each cell and transmits data only in the First Time Slot (TSO) of certain GSM frames.
- Unlike Traffic Channels (TCHs) which are duplex, BCCHs only use the forward link.
- Just as the Forward Control Channel (FCCH) in AMPS is used as a beacon for all nearby mobiles to camp onto.
- The BCH provides synchronization for all mobiles within the cell and is occasionally monitored by mobiles in neighbouring cells so that received power and Mobile Assisted Hand-off (MAHO) decisions may be cut-off-cell users.
- Although BCH data is transmitted in TSO, the other seven time slots in a GSM frame for that same ARFCN are available for traffic channel data, DCCH data, or filled with dummy bursts.

- Furthermore, all eight time slots on all other ARFCNs within the cell are available for TCH or DCCH data.
- The BCH is supported by three separate channels which are given access to TSO during various frames of the 51 frame sequence.

**(a) Broadcast Control Channel (BCCH) :**

- The BCCH is a forward control channel that is used to broadcast information, such as cell and network identity and operating characteristics of the cell (current control channel structure, channel availability and congestion).
- The BCCH also broadcast a list of channels that are currently in use within the cell.
- It should be noted that TSO contains BCCH data during specific frames and contains other broadcast channels (BCCH) and Synchronization Channel (SCH).

**(b) Frequency Correction Channel (FCCH) :**

- The FCCH is a special data burst which occupies TSO for the very first GSM frame (frame 0) and is repeated very ten frames within a control channel multiframe.
- The FCCH allows each subscriber unit to synchronize its internal frequency standard (local oscillator) to the exact frequency of the base station.

**(c) Synchronization Control Channel (SCCH) :**

- The SCH is broadcast in TSO of the frame immediately following the FCCH frame and is used to identify the serving base station while allowing each mobile to frame synchronize with the base station.
- The Frame Number (FN), which ranges from 0 to 2, 715, 647 is sent with the *Base Station Identity Code* (BSIC) during the SCH burst.
- The BSIC is uniquely assigned to each BST in a GSM system.
- Since a mobile may be as far as 30 km away from a serving base station, it is often necessary to adjust the timing of a particular mobile user such that the received signal at the base station is synchronized with the base station clock.
- The base station issues coarse timing advancement commands to the mobile stations over the SCH, as well.
- The SCH is transmitted once every ten frames within the control channel multiframe.

**(iii) Common Control Channels (CCCHs) :**

- **A Base Transceiver Station (BTS), when granting access to a Mobile Station (MS) so that MS can use either SDCCH or transfer channel uses a channel called Access Grant Channel (AGCH).**
- After the access is granted, the call set up or call forwarding can take place.
- The control channel used for such purposes is called a CCCH.
- On the Broadcast Channel (BCH) ARFCN, the common control channels occupy TSO of every GSM frame that is not otherwise used by the broadcast channel or the idle frame.
- The idle frame CCCH are the most commonly used control channels and are used to page specific subscribers, assign signaling channels to specific uses and receive mobile requests for service.
- The CCCH consists of three different channels namely, (a) Paging channel, (b) Random access channel, and (c) Access grant channel. These channels are as described below :

  **(a) Paging Channel (PCH) :**

  - **When call forwarding information is transmitted from the base transceiver station, the CCCH is called Paging Channel (PCH).**
  - An example of paging is transmission of information to a select target mobile station for example, the identity of the caller of an incoming call to the mobile station to which the call is to be forwarded.
  - The paging channel provides paging signals from the base station to all mobiles in the cell and notifies a specific mobile of an incoming call which originates from the PSTN.

- The paging channel transmits the IMSI of the subscriber, alongwith a request for acknowledgement from unit on the Random Access Channel (RACH).
- Alternatively, the paging channel may be used to provide cell broadcast ASCIT text messages to all subscribers, as part of the SMS feature of GSM.

**(b) Random Access Channel (RACH) :**

- **When call set up requirements are transmitted from the mobile station, the common control channel is called a Random Access Control Channel (RACH).**
- The RACH data burst format is as follows : during 577 μs in the place of sequence (H, user data S, TR, S, user data and T), a 145-bit sequence is modified as eight H bits, 41 synchronization bits, 36 bits user data and 3T bits (total 88 bits).
- The guard-space time intervals are now equal to $\frac{(68.25 \times 3.692)}{2}$ = 126 μs before H and after T bits.
- The RACH is reverse link channel used by a subscriber unit to acknowledge a page from the paging channel and is also used by mobiles to originate a call.
- It uses a slotted ALOHA access scheme.
- All mobiles must request access or respond to a paging channel alert within TSO of a GSM frame.
- At the base transceiver station, every frame (even the idle frame) will accept RACH transmission from mobiles during TSO.
- In establishing service, the GSM base station must respond to the RACH transmission by allocating a channel and assigning a Standalone Dedicated Control Channel (SDCCH) for signaling during a call.
- This connection is confirmed by the base station over the Access Grant Channel (AGCH).

**(c) Access Grant Channel (AGCH) :**

- **When access granting information is transmitted from the base transceiver, the common control channel is called Access Grant Channel (AGCH).**
- The access grant channel is used by the base station to provide forward link communication to the mobile, and carries data which instructs the mobile to operate in a particular physical channel (TS and ARFCN) with a particular dedicated control channel.
- It is the final Common Control Channel (CCCH) message sent by the base station before a subscriber is moved off the control channel.
- It is used by the base station to respond to a random access channel sent by a mobile station in a previous common control channel frame.

### 5.13.7 Reverse Traffic Channels

- Reverse traffic channels are used for transmitting voice or data traffic from a mobile to the base station.
- Each of these channels is paired with a corresponding forward traffic channel signaling information during a call is again transmitted in **blank-and-burst** or **dim-and-burst** mode.
- The access channel operates at 4800 b/s. Some key features for the CDMA reverse link channels are as follows :
  1. Voice information is digitized using a QCELP variable rate codes.
  2. All information bits are protected using rate 1/3 convolution code and block inter-leaving to achieve implicit time diversity.
  3. A stronger rate 1/3 convolution code is used to compensate for the absence of a pilot channel (which generally assists in coherent direction).
  4. The reverse channels are tightly power controlled by the base station to minimize interference.
  5. Each mobile station is assigned a unique PN spreading sequence.

## 5.14 GSM FRAME STRUCTURE

### 5.14.1 Introduction

- The data frames and time slots within 2G GSM are organized in a logical manner so that the system understands, when particular types of data are to be transmitted.
- Having the GSM frame structure enables the data be organized in a logical fashion so that the system is able to handle the data correctly. This includes not only the voice data, but also the important signaling information as well.
- The GSM frame structure provides the basis for various physical channels used within GSM, and accordingly it is at the heart of the overall system.

### 5.14.2 Definition

- **Frame structure is the division of defined length of digital information into different fields (information parts).**
- A GSM frame is 4.615 ms and it is composed of 8 time-slots (numbered 0 through 7). During voice communication, one user is typically assigned to each time-slot within a frame. The GSM system also combines frames to form multiframes.

### 5.14.3 Basic Principle

- The basic element in the GSM frame structure is the frame itself. This comprises 8 states, each used for different users, within the TDMA system. The time-slots for transmission and reception for a given mobile are offset in time, so the mobile does not transmit and receive at the same time.
- The basic GSM frame defines the structure upon which all the timing and structure of the GSM messaging and signaling is based. **The fundamental unit of time is called a burst** period and it lasts for approximately 0.577 ms (= 15/26 ms).
- Eight of these burst periods are grouped into a TDMA frame. This lasts for approximately 4.615 ms (= 120/26 ms) and it forms the basic unit for the definition of logical channels. One physical channel is one burst period allocated in each TDMA frame.
- In simplest terms the base station transmits two types of channel, namely traffic and control. Accordingly the channel structure is organized into two different types of frame, one for traffic on the more traffic carrier frequency and the other for the control on the Beacon frequency.

### 5.14.4 Multiframes

- Multiframes are frames that are grouped or linked together to perform specific functions. Multiframes on the GSM system use established schedules for specific purposes, such as coordinating with frequency hopping patterns.

### 5.14.5 Types

The types of multiframes used in GSM system are as under :

1. Traffic multiframe structures.
2. Control multiframe structures.
3. Super frame multiframe structures.
4. Hyper multiframe structures.

### 5.14.6 Frame Structure

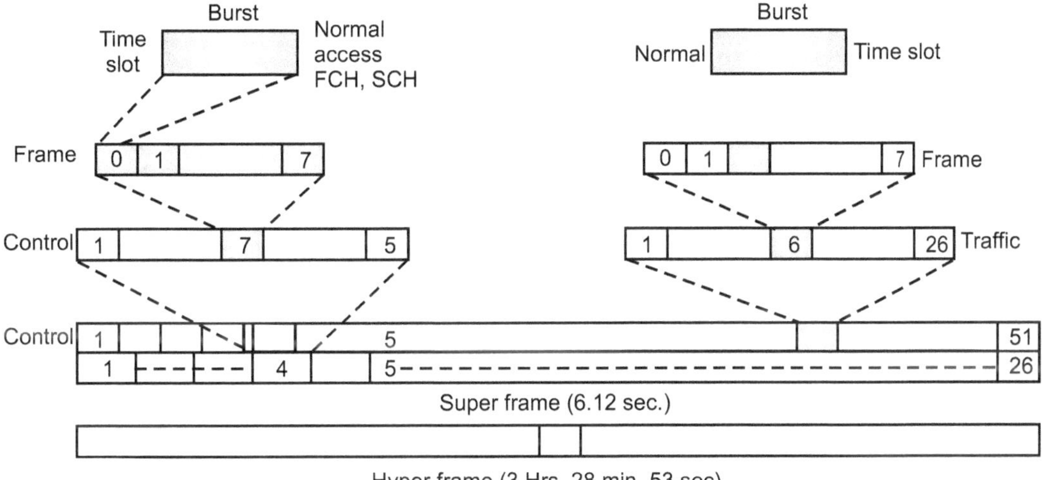


**Fig. 5.6 : GSM frame structure**

- Fig. 5.6 shows different types of GSM frame and multiframe structures. This diagram shows that a single GSM frame is composed of 8 time-slots. When a radio channel is used to provide a control channel, time-slot 0 and the other time state are used for traffic channels.
- 51 frames are grouped together to form control multiframes for the control channel. 26 frames are grouped together to form traffic multiframes for the traffic channels. Superframes are the composition of 26 control multiframes or 52 traffic multiframes to provide a common time period of 6.12 seconds. 2648 superframes are grouped together to form a hyperframe. A hyperframe has the longest time period in the GSM system of 3 hours, 28 minutes and 53 seconds.

## 5.15 SPEECH PROCESSING

### 5.15.1 Brief History

- Early attempts at speech processing and recognition were primarily focused on understanding a handful of simple phonetic elements such as vowels. In 1992, three researchers at Bell Laboratories, Stephen Balashek, R. Biddulph, and K. H. Davis developed a system that could recognize digit spoken by a single speaker.
- Linear Predictive Coding (LPC), a speech processing alongwith, was first proposed by Fumidata Itak ura of Nagoya University and Shuzo Saito of MTT in 1966. Further developments in LPC technology were made by Bishnu S. Atal and Mantred R. Schrosder at Bell Laboratories during the 1970s. LPC was the basis for voice-over - IP (Vo-IP) technology, as well speech synthesizer chips, such as Texas Instruments LPC speech chips used in the speech and spell toys from 1978.
- One of the first commercially available speech recognition products was Dragon Dictate, released in 1990. In 1992, technology developed by Lawrence Robiner and others at Bell Laboratories was used by AT&T in their Voice Recognition Call Processing (VRCP) in service to route calls without a human operator.
- By this print, the vocabulary of these systems was longer than the average human vocabulary. By the early's 2000s, the dominant speech processing strategy started to shift away from Hidden Markov Models (HMMs) towards more modern Neural Networks and deep learning.

### 5.15.2 Introduction

- Speech may be more intuitive way of accessing information, controlling things and communicating, but there may be viable things and communicating, but there may be way to interact with your computer. *Speech is hands-free, eyes-free, fast and intuitive.*

- Applications can be interactive or non-interactive. Conversational human machine interfaces would be considered interactive and fall into quadrant A. Monitoring speech communicationsless so and would fall into quadrant B. Additionally, we can consider real-time versus non-real-time. Live speech transcription falls into quadrant B, but speech data fining into quadrant C.

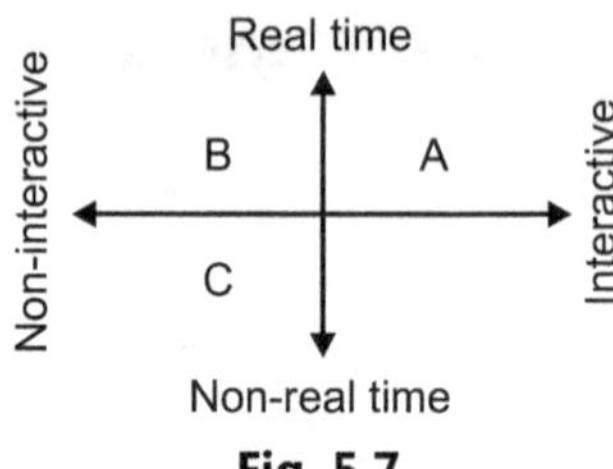

**Fig. 5.7**

- Applications of speech technology can be grouped into communication and control, communications, which would fall into quadrant A. Information which would cross quadrants A, B and C, Intelligence in quadrants B and C and processing in the remaining quadrant. Tasks get increasingly harder the further away from the centre of the chart it becomes.

### 5.15.3 Definition

- *Speech processing is the study of speech signals and the processing methods of these signals, and also as the intersection of digital signal processing and natural language processing.*
- Aspects of speech processing include the acquisition, manipulation, storage, transfer and output of speech signals. The input is called *speech recognition* and the output is called *speech synthesis*. The speech signals are usually processed in digital in a digital representation. So, the speech processing can be regarded as a special case of *digital signal processing*, applied to the speech signal.
- *Speech is continuous, variable, ambiguous, rich in information, multi-modal, contaminated, disfluent and multi-disciplinary.*

### 5.15.4 Speech

- Speech is the most sophisticated behaviour of the most complex organism in the known universe. Understanding how speech works is much more challenging than rocket science.
- Within any given language, groups of individuals use local **didets** (local words) and **accent** (different sounds), which reflect regional or social situations.
- Even within a given language group, different individuals sound different due to inter-speaker variation (age, gender, physical characteristics e.g., vocal tract size and social habits) and intra-speaker variation (physio-social, psychological and external factors.
- These external factors include noise, leading to hyper-articulation or to the Lambard effect (varying amplitude to account for noises in the environment), vibrations, what the tasks is, who the listener to emotion, cognitive load and alcohol or drugs.
- Speech is not just acoustic signal, but it is also visual. Listeners can derive a large benefits from seeing a speaker's lipo-equivalent to a 12 dB SNR in a noisy environment. This visual information can even override the acoustic information.
- Speech is also contaminated - every day speech, which occurs in normal environments include other voices, competing sounds, noise and distortion, and reverberation.
- Speech is also diffluent. Normal spontaneous speech contains false-starts, repeats, filled pauses (ums and ahas) and overlaps. Spoken utterances are not as well-formed as written sentences. These so called disfluencies make speech easy for a human to produce and understand.
- Finally, speech is multi-disciplinary. The different topics of speech are built from aspects of engineering, logistics and psychology.

### 5.15.5 Speech Structure

- Speech is a complex phenomenon. It is a continuous audio stream, where rather stable states mix dynamically changed data. In this sequence of states, one can define more or less similar classes of sounds or phones words are understand to be built of phones, but this is certainly not true. The acoustic properties of a waveform corresponding to a phone can vary greatly depending on many factors-phone context, speaker, style of speech and so on. The so called coarticulation makes phones sound very different from their *'canonical'* representation. Next, since transitions between words are more informative than stable regions, developers often talk about *diphones* - parts of phones between two consecutive phones. Sometimes developers talk about subphonetic units - parts of phones between two consecutive phones. Sometimes developers talk about subphonetic units - different substates of a phone often 3 or more regions of a different nature can be found.

## 5.15.6 Techniques

1. Dynamic time warping.
2. Hidden Markev models.
3. Artificial neural networks.

## 5.15.7 Applications

1. Interactive voice systems.
2. Virtual assistants.
3. Voice identification.
4. Emotion recognition.
5. Call center-automation.
6. Robotics.
7. Digital speech coding.
8. Spoken language dialog systems.
9. Text-to-speech synthesis.
10. Automatic speech recognition.

## 5.15.8 Speech Processing Flow in GSM

- During a call, speech signal has to be digitized, coded, formatted, modulated and then transmitted on wireless media. The entire signal processing flow is discussed as follows. GSM physical layer is nothing but the modules through which speech will pass through before they are transmitted in the air.

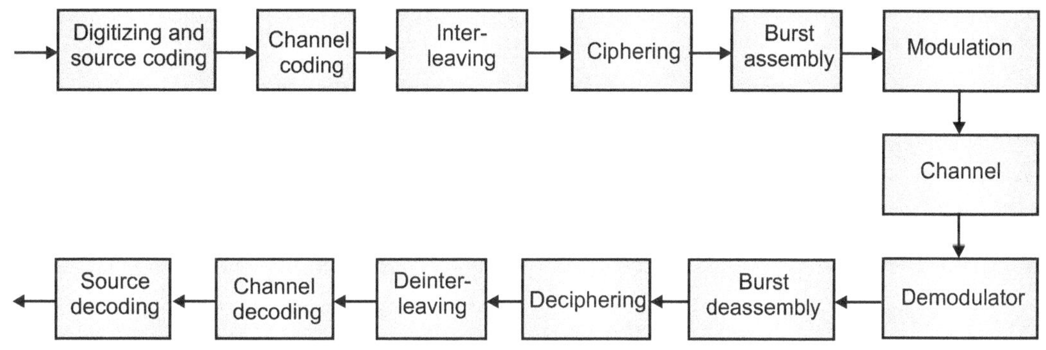


**Fig. 5.8 : GSM speech processing flow**

- The modules of GSM speech processing flow are as under :
    1. Speech coding
    2. Channel coding
    3. Inter-leaving
    4. Ciphering
    5. Burst assembly
    6. Modulation
- Speech coding block uses 13 kbps RELP (Residuary Excited Linear Predictive) coder. Channel coding block uses convolution coding of rate ½ with constraint length of 5. Inter-leaving block does diagonal inter-leaving after 456 encoded bits in 20 ms duration are broken into 57 bits sub-blocks.
- There will be about total 8 sub-blocks of 57 bits each. Ciphering block uses $A_3$ and $A_5$ encryption algorithms. Encryption is changed call by call to enhance privacy. Burst assembly block frames the bursts as required by the GSM frame structure. The same is modulated and Gaussian filtered.
- Modulation block minimizes the occupied bandwidth using GMSK modulation with BT of 0.3.

## 5.16 3G TECHNOLOGY

### 5.16.1 Introduction

- 3G wireless technology refers to the third generation technology of wireless networks. It is an enhanced version of 1G as well as of the 2G technology.
- The 3G technology is also known as Universal Mobile Telecommunication System (UMTS) providing wireless access to the data and information to the users.
- It was successfully launched and introduced to the millions of the consumers all over the globe. It was first released in Japan in October, 2001.
- The 3G phone supporting this technology was designed for the users to have the ability to surf the internet, watch movies, download songs and videos, and view pictures of other individuals that are in contact with. It works at the speed of 2 Mbps.
- The speed efficiency and connectivity is much better than the previous 1G and 2G technologies.
- Users can perform all the functions on 3G phone with the enhanced speed, efficiency and connectivity as it offers a faster connectivity as per the users requirement including a much better video call facility, clearer communication and faster speed when two parties are using the same 3G technology. The transferring of the data through wireless 3G technology is regardless of the time and location.

### 5.16.2 Need

- The need for the 3G wireless technology in accordance as per the need for the much more data speed and the requirement for more efficient applications in accordance to overcome the disadvantages or the lack of functionality provided by 1G and 2G technologies. The speed, efficiency and connectivity of 3G is much better than the previous technologies. The 3G technology has various advantages such as quicker connectivity, music, entertainment with better quality and faster access of the internet.

### 5.16.3 Key Features

1. It has multi-megabit internet access.
2. It has voice activated cells.
3. It has unparallel network capacity.
4. It obiguitors always on access.
5. It has communication using Voice Over Internet Protocol (VOIP).

### 5.16.4 Advantages

1. Several times higher data sped.
2. Enhanced audio and video streaming.
3. Video conferencing support.
4. Web and WAP browsing at higher speed.
5. IPTN (TV through the internet) support.
6. Much better speed, efficiency and ability.
7. Better entertainment applications.
8. Better video call facility.
9. Clearer communication.
10. Quicker connectivity.
11. Faster access of the internet.
12. Better mobility and remote connectivity.
13. Better storage and sharing.
14. Automation

### 5.16.5 Disadvantages

1. High prices to help cover the initial demand and fixed cost.
2. Needs more towers and high density requires towers to be closer together.
3. Availability of consumers.
4. One of the major disadvantage is that it would be a big risk for entrepreneurs to take because the costs are high and the start is slow.
5. Requires compatible handset.
6. High power consumption.
7. High upgrading cost.

### 5.16.6 Applications

1. Global Positioning System (GPS).
2. Location-based services.
3. Mobile TV.
4. Telemedicine.
5. Video conferencing.
6. Video on demand.
7. It is used in personal applications.
8. It is used in business applications.
9. It is used in productivity applications.
10. It is used in control applications.
11. It is used in communication applications.

### 5.16.7 Standard Systems

1. W-CDMA (UMTS).
2. CDMA - 2000.
3. IMT - 2000.
4. TD-S CDMA.

## 5.17 UMTS STANDARDS

### 5.17.1 Introduction

- **UMTS stands for Universe of Mobile Telecommunications System.** It is being developed by RACE (Rand in Advanced Communication technologies in Europe) as the Third Generation (3G) wireless system. To handle a mixed range of traffic, a mixed cell layout (shown in Fig. 4.4) that would consist of microcells overlaid on microcells and picocells is one of the architecture plans being considered.
- This type of network distributes the traffic with the local traffic operating on the microcells and picocells, while the high mobile traffic is operated on the microcells. Thus, reducing the number of hand-offs required for the fast moving traffic. UMTS uses a totally different radio interface based around the use of direct sequence spread spectrum as CDMA (code division multiple access).
- Although 3G UMTS uses a totally different radio access standards, the core network is the same as that used for GPRS and EDGE to carry separate circuit switched voice and packet data. UMTS uses a wideband version of CDMA occupying a 5 MHz wide channel. Being wider than its competition CDMA 2000 which only used 1.25 MHz channel, the modulation scheme was known as wideband CDMA or W-CDMA/WCDMA. This name was often used to refer the whole system.

### 5.17.2 Definition

- UMTS using WCDMA technology is a third generation (3G) mobile communications system run under the category of 3 GRP providing mobile data connecting and circuit switched voice.
- UMTS is the 3G successor to the GSM family of standards including GPRS and EDGE.
- *UMTS is a third generation (3G) broadband, packet based transmission of text, digitized voice, video and multimedia at data rates upto 2 Mbps.*
- *UMTS is a system that is capable of providing a variety of mobile services to a wide range of GSM standards.*

### 5.17.3 Switched Elements

1. **Circuit Switched Elements :**

- These elements are primarily based on the GSM network entities and carry data in a circuit switched manner, e.g. a permanent channel for the direction of the call.
- The circuit switched elements of the UMTS core network architecture includes the following network entities.
  - **(a) Mobile Switching Centre (MSC) :** This is essentially the same as that within GSM, and it manages the circuit switched calls underway.
  - **(b) Gateway MSC (GMSC) :** This is effectively the interface to the external networks.

2. **Packet Switched Elements :**

- These network entities are designed to carry packet data. This enables much higher network usage as the capacity can be shared and data is carried as packets, which are routed according to their destination. The packet switched elements of the 3G UMTS core network architecture include the following :
  - **(a) Serving GPRS Support Node (SGSN) :** As the name implies, this entity was first developed when GPRS was introduced and its use has been carried over into the UMTS network architecture. The SGSN provides a number of functions within the UMTS network architecture.
  - **(b) Gateway GPRS Support Node (GGSN) :** Like the SGSN, this entity was also first introduced into the GPRS network. The GGSN is the central element within the UMTS packet switched network.
- It handles inter-working between the UMTS packet switched network and external packet switched network. In operation when the GGSN receives the data addressed to a specific user, it checks if the user is active and then forwards the data to the SGSN serving the particular UE (user equipments).

3. **Shared Elements :**

- The shared elements of the 3G UMTS core network architecture include the following network entities : (i) Home location register (HLR), (ii) Equipment identity register (EIR), (iii) Authentication Centre (ATC).

## 5.17.4 Architecture

- With the changes from 2G to 3G, the emphasis for the systems changed from focus on mobile voice communications to mobile data and general connectivity.
- The foundation for the UMTS network had been set in place when GSM was launched. This provided the basic access elements as well as circuit switched voice. The additional network entities to be added it was the combination of these two network elements that provided the basis for 3G UMTS network architecture.
- The architecture of UMTS network is as shown in Fig. 5.9. The UMTS core network may be split into two different areas : (i) Circuit switched elements, (ii) Packet switched elements.

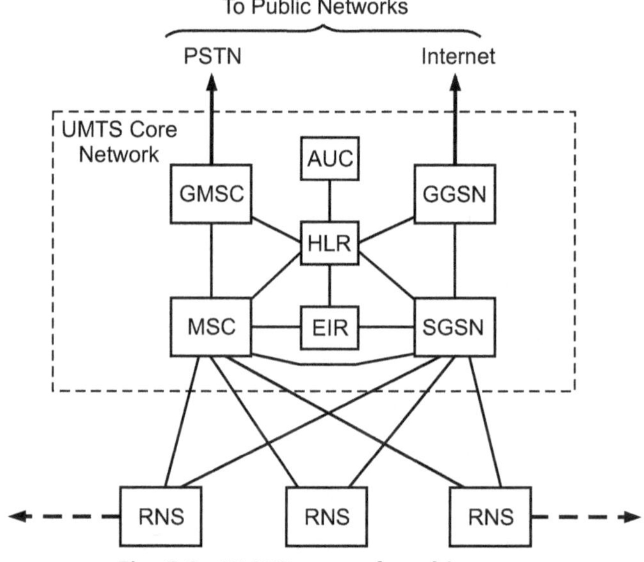


**Fig. 5.9 : UMTS network architecture**

## 5.17.5 Air-Interface Specifications

| | | |
|---|---|---|
| 1. Frequency spectrum | : | Uplink - 1920 to 1980 MHz<br>Downlink - 2110 to 2170 MHz |
| 2. Channel bandwidth | : | 5 MHz |
| 3. Chip rate | : | 3.84 Mbps |
| 4. Duplexing technique | : | FDD and TDD modes |
| 5. Modulation scheme | : | Direct sequence CDMA and QPSK |
| 6. Frame Length | : | 10 ms frame with 15 time slots |
| 7. Coding technique | : | Orthogonal variable spreading factor (OVSF) |
| 8. Service type | : | Multi-rate and multi-service |

### 5.17.6 UMTS Security Procedure

- The UMTS is one of the 3G mobile cellular communication systems. It builds on the success of the 2G GSM system. One of the factors in the success of GSM has been its security features. New services introduced in UMTS require new security features to protect them.
- In addition, certain real and perceived shortcomings of GSM security need to be addressed in UMTS.

1. **Entity Authentication :**

- UMTS provides mutual authentication between the UMTS-subscriber, represented by a smart card castle application known as USIM (Universal Subscriber Identity Module) and the network in the following sense- *Subscriber authentication.*
- The following sense network collaborates the identity of the subscriber and 'Network authentication'. The subscriber collaborates that he is connected to a serving network that is authorized by the subscribers home network to provide.

2. **Signaling Data Integrity and Origin Authentication :**

- **Integrity algorithm agreement :** The mobile station and the serving network can securely negotiable the integrity algorithm that they use.
- **Integrity key agreement :** The mobile and the network agree on an integrity key that they may use subsequently, this provides entity authentication.

3. **User Traffic Confidentiality :**

- The term **'Network domain security'** in the 3G covers security of the communication between network elements. In particular, the mobile station is not affected by network domain security.
- The two communicating network elements may both be in the same network administrated by a mobile operator or they may belong to two different networks.

4. **MAPSEC :**

- The basic idea of MAPSEC can be described as follows :

  The plain text MAP message is encrypted and the result is put into a **'container'** in another MAP message.
- As the same time, a cryptographic check sum i.e. a message authentication code covering the original message, it included in the new MAP message.
- To be able to use encryption and message authentication codes, keys are needed. MAPSEC has borrowed the nation of a Security Association (SA) from IPSEC.

5. **IP Multimedia System Security :**

- The IP multimedia sub-system is a core network sub-system within UMTS. It is based on the use of the Session Initiation Protocol (SIP) 2G to initiate, terminate and modify multimedia sessions such as voice calls, video conferences, streaming and chat.
- SIP is specified by the Internet Engineering Task Force (IETF) 27. IMS also uses the IETF Session Description Protocol (SDP) 28 to specify the session parameters and to negotiate the codes to be used.
- SIP runs on top of different IP transport protocols such as User Datagram Protocol (UDP) and the Transmission Control Protocol (TCP).
- A 3G IMS subscriber has one IP multimedia private identity (IMPI) and at least one IP multimedia public identity (IMPI). To participate in multimedia sessions, an IMS subscriber must register in multimedia sessions, an IMS subscriber must register at least on IMPS with the IMS. The private identity is used only for authentication purposes.

## 5.18 CDMA-2000 STANDARDS

### 5.18.1 Introduction

- The basic structure of 3G CDMA-2000 is same as that of 2G CDMA system with a channel bandwidth of 1.25 MHz per radio channel. The 3G CDMA-2000 standards are based on the original IS-95, IS-95, IS-95A (CDMA-1) and 2.5G IS-95B air interface standards. It provides high data rate internet access capabilities in a gradual manner within the existing systems.
- It also provides the backward compatibility with the existing CDMA-1 or IS-95B equipments.

### 5.18.2 Definition

- The 3G CDMA-2000 is a set of Third Generation (3G) standards and it is registered under the trademark of Telecommunication Industry Association (TIA) of U.S.A. It is an upgradation of the existing 2G and 2.5G CDMA technology and it can support much higher data rates than that of 2G and 2.5G systems.
- It has improved capabilities over 3G W-CDMA at each cell without changing the base stations entirely. In the first 3G CDMA-2000 air interface, it was proposed to use a single 1.25 MHz radio channel. This is called 3G CDMA-2000 $1_X$.
- The 3G CDMA-2000 $1_x$ can support an instantaneous data rate of upto 307 kbps for a user in packet mode and yields a typical throughput rate of about 144 kbps per user. The number of voice users that can be supported by 3G CDMA-2000 is almost twice that of the 2G CDMA system.
- The 3G CDMA-2000 $1_x$ can provide the subscriber unit with upto twice the stand by time for longer battery life. The 3G CDMA-2000 is being developed for both FDD (for mobile radio) and TDD (in building cordless) applications. The improvement in 3G CDMA-2000 $1_x$ over 2G and 2.5G CDMA systems are gained through the use of rapidly adaptable baseband signaling rates and ciphering rates for each user.
- No additional RF equipment is needed or the RF system components of the base station need not be changed to enhance the performance but the changes in the software or in baseband hardware are to be made. The 3G CDMA-2000 $1_x$ EV is the advanced version of CDMA for providing a high data rate (HDR) packet standard to be overlaid upon the existing IS-95, IS-95B and CDMA-2000 networks.
- In 3G CDMA-2000 $1_x$ EV, the user has an option of installing the radio channels with data only (CDMA-2000 $1_x$ EV - DO) or with data and voice (CDMA-2000 $1_x$ EV - DV). Using 3G CDMA-2000 $1_x$ EV system, individual 1.25 MHz channel can be installed in the base station of CDMA to provide high speed packet data access within the selected cells. If 3G CDMA-2000 $1_x$ EV - DO option is used, then it supports the data rates in excess of 2.4 Mbps. The 3G CDMA-2000 $1_x$ EV - DV system supports the data as well as voice transmission. It can offer data rates upto 144 kbps and it can support with about twice the number of voice channels as IS-95B.

### 5.18.3 Key Features

Some of the important features of 3G CDMA-2000 are as under :

1. It is more robust for multipath delays.
2. It provides higher immunity towards frequency selective fading.
3. It has very high packet data rates.
4. It has high radio channel bandwidth of 1.25 MHz.
5. It supports almost twice the number of voice users that of 2G CDMA system.
6. It provides high speed packet data access within the selected cell.
7. It is possible to upgrade the 2G CDMA system to CDMA-2000 system using the same spectrum bandwidth, RF equipment and air interface framework at each base station.
8. It has backward compatibility with the 2G CDMA or IS-95B systems.

9. It has global seamless connectivity (roaming).
10. It provides the subscriber with upto twice the stand by time for longer battery life.
11. It has a wide range of telecommunication services such as voice, data, multimedia, internet etc.
12. It can operate in multiple radio environments such as cellular, cordless, satellite, LAN etc.

### 5.18.4 Description

The operation of 3G CDMA-2000 is discussed as under :

1. **External synchronization mechanism :**
   - The 3G CDMA-2000 base stations (BTSs) use an external synchronization mechanism satellite based GPS. The adjacent cells can use the same frequency but must use distinct phase angles. Some period of time is required for synchronization among the adjacent base stations in asynchronous stations in 3G W-CDMA.
   - Therefore synchronizing processes in 3G W-CDMA stations have a certain latency period. The latency between base stations of adjacent cells causes call drops in 3G W-CDMA, even when a soft handoff is performed. The external synchronization among base stations in 3G CDMA-2000 reduces latency related call drops.
2. **Dedicated and common channels :**
   - The data-link layer MAC (Message Authentication Code) protocols between MS and BTS support two types of channels, dedicated channels and common channels for common control. The reverse access channel and reverse pilot channel are the only common control channels present during the access set-up phase.
   - The frame structure is 172 users data bits, 12 CRC (Cyclic Redundancy Check) bits, and 8 tail as 0s for a 4.8 kbps access channel. There is a preamble before a message. The access is random and uses the slotted ALOHA protocol. The user data is not transmitted through the access channel, but through dedicated channels with rate sets RS1 or RS2. On putting the user channel data in the 20 ms time-slots, the rate becomes 307.2 kbps so that after a four chip Walsh code (spread factor of 4), the chipping rates become 1.2288 Mchip/s.
3. **Short and long data packets :**
   - The low data services employ one fundamental mode and high data rate services and multimedia services employ a greater number of dedicated channels. A pilot signal is a reference signal. The mobile station uses the continuously transmitting, code-divided, dedicated pilot for uplink.
   - The base transceiver station uses the code-divided common pilot and dedicated/common auxiliary pilots for downlink. The code divided and dedicated pilot helps in coherent dedication of the reference signal. The BTS pilot channel multiplexes the power control and a control bit called *ensure indicator bit.*
4. **Physical channels :**
   - The frame length modulation is QSPK for downlink and BPSK for up link. It gives frames of 5, 10, 20, 40 and 80 ms duration as per the packet size.
   - The short data bursts use the slotted ALOHA protocol and are transmitted at variable power. The power level is enhanced after an unsuccessful access.
5. **Multi-rate transmission of signals :**
   - The 3G CDMA-2000 supports several different types of physical channels. The chip rates are $n$x 1.2288 Mchip/s, where n is a positive integer, $n > 1$ facilitates multi-rate or single rate data transmission at higher rates, which are n times 1.2288 Mbps.
   - The supplementary channels for reverse link (uplink) use turbo codes. Other physical channels use 1/4 convolution encoding. The signals are scrambled with long code sequences before chipping with orthogonal codes. This is followed by short code spreading of the carrier channels.

- The long codes in 3G CDMA-2000 use the same length M-sequences but are used for channel data divided in two different phases (I and Q) for each channel before scrambling. A Third Generation (3G) system must support multi-rate transmission of signals. An RS2 signal can be punctured to reduce data rates.
- The low data-rate signals of upto 307.2 kbps, variable length Walsh codes support both orthogonality as well as variable data rate for a physical channel. The spread factor can vary from 4 to 256 depending upon the data rates in 3G CDMA-2000. The IS-95 spread factor is constant at 64. The downlink uses Walsh codes or quasi-orthogonal codes and uplink uses Walsh codes. The orthogonal coding for channelization is asymmetrical in uplink and downlink.

6. **Multi-rate data encoding in traffic channels :**
   - The IS-95 CDMA-one user data has different data rates. The rate set RS1 transmits at 9.6, 4.8, 2.4 or 1.2 kbps. An optional rate set RS2 transmits at 14.4, 7.2, 3.6 or 1.8 kbps. It requires a different type of convolution encoding and error encoding depending on considerations of service quality. The low or high service quality means high or low bit error rates, respectively.
   - As an example, multi-encoded rate signals are systems that transmit at variable data rates after convolution coding. In IS-95, the RS1 and RS2 data rate symbols are matched, repeated and interleaved so that before chipping, the symbol transmission rate is constant at 19.2 k symbol/s.

7. **Downlink and uplink modulations of spreading signals (Pilots) :**
   - The spreading signal (pilots) modulation is balanced QPSK modulation for downlink and dual channel QPSK modulation for uplink. The modulation of radio-carrier frequency is asymmetrical for uplink and downlink.

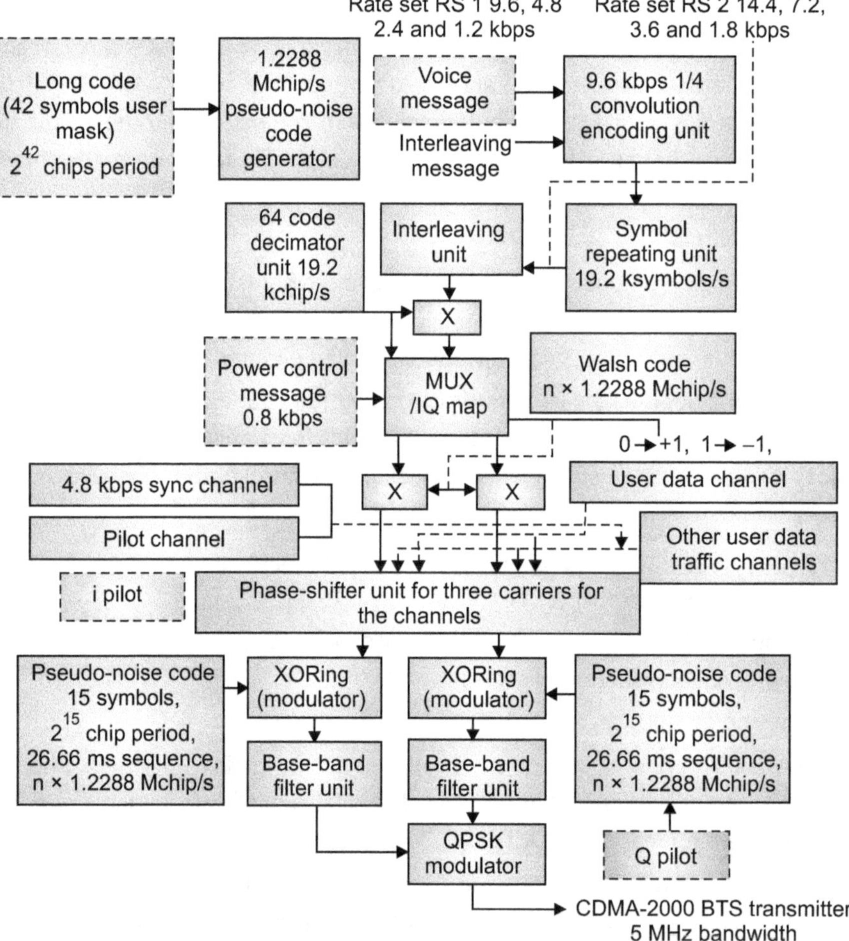


**Fig. 5.10 : Processing units used in the three-carrier CDMA-2000 3x standard**

**8. Multi-carrier rate support for transmission of signals :**

- The use of interleaving and variable spread factor enables support of the data rates upto 386 kbps. When high data-rate signals, for example, video signals are to be transmitted, then multi-carrier transmission can be used. We can use *n* carriers, where *n* = 1, 3, 6, 9 or 12 in 3G CDMA-2000 1x, 3x and so on. Similarly, chipping rates of *n*x 1.2288 Mchip/s, where *n* = 1, 3, 6, 9 or 12 can be used.
- The bandwidth for 3G CDMA-2000 3x is 5 MHz. While uplink uses DS-CDMA, downlink can use either multi-chip rate CDMA. Rate of *n*x 1.2288 Mchip/s, where *n* is the number of carriers or DS-CDMA. Assuming that there is no mapping of 1s and 0s. In 3G CDMA-2000 systems, there are three carriers of 1.25 MHz each for *n* = 3 and all carriers have a separate code for each channel.
- It has various channels, i.e., user data channels, user traffic channels, synchronization channels and pilot channels. Therefore each of the *n* channels is transmitted as a separate carrier and is identified and separated at the receiver. The multi-carrier rates can be adopted in 3G CDMA-2000.
- It means that when *n*-channels data is to be transmitted, then each carrier uses 1.2288 Mchip/s for transmission. When single channel data or time division multiplexed data of different channels is to be transmitted at high data rates, then a single carrier uses *n*x 1.2288 Mchip/s for chipping. One option for employing chipping rates of 1.2288 Mchip/s when transmission is at higher ratio (= *n*x 1.2288 Mchip/s) is as follows : one channel is transmitted in a time-slot by time division multiplexing and the same chipping rate is used for that channel.
- Another option is that only a single carrier be transmitted at an instant but at high transmission rates of *n*x 1.2288 Mbps and using 1.2288 Mchip/s. The W-CMDA uses the second option (3.84 Mchip/s). The multiplexing in time and code space can be used for sending data from different channels and for sending data from different traffic channels. The 3G CDMA-2000 uses the first option and three or more carriers (each with a chipping rate of 1.2288 Mchip/s) can be transmitted by multiplexing in time-space.

## 5.18.5 Processing Units

- Fig. 5.10 shows the processing units used in the 3G CDMA-2000 standard for ower control messages and channels for data sync., pilot and traffic base-band transmission and use of a MUX/IQ unit, long code and mask, $PN_a$, $PN_i$ and Walsh codes.
- In IS-95, the data channel is divided into I (in-phase) and Q (quadrature) components after chipping with Walsh codes. In CDMA-2000, data channel is divided before scrambling.
- The main difference with respect to the IS-95 processing unit is that in IS-95, the data channel divided into I (in-phase) and Q (quadrature) components after scrambling with Walsh code and in CDMA-2000, the MUX/IQ (multiplexer and signal mapping unit) first divides the signal into the I and Q from other user traffic, user data, pilot and synchronization channels are multiplexed and passed through PN short-code spreaders and base-band filters.
- There are three sets of chips after multiplexing when the chipping rate is 3 times 1.2288 Mchip/s. Each chip has two components I and Q. After the PN short code encoding with I and Q pilots and base-band filtering of I and Q components, a balance QPSK modulator modulates the signal for downlink and a dual channel QPSK modulator modulates it for uplink.

## 5.18.6 Phase Shifter Unit

- The phase shifter unit for three carriers for the channels is shown in Fig. 5.11. It shows the phase shifted three-carrier transmission of carrier channels A, B and C by multiplexing in time space. The code 0110101001 has been used as an example in Fig. 5.11.

- The actual code must be an orthogonal code and must have a code length per user symbol as per transmission channel requirements.
- Also 0s are transmitted as +1s and 1s as –1s using signal mapping as per 3G PP specifications.
- All channels are chipped using the same orthogonal code and scrambled using the same PN long codes, but the chipping instants of each carrier channel are shifted by (1/3) × 814 *ns* (phase angle change of 120°), so that the combined chipping rate is 1.2288 × 3 = 3.6814 Mchip/s.

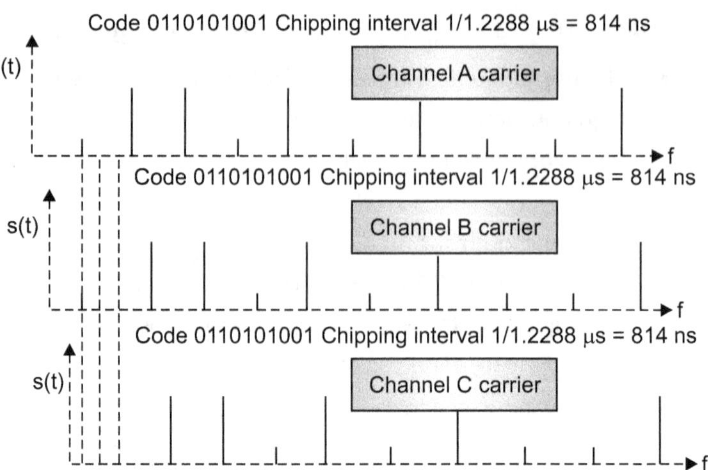


**Fig. 5.11 : Three-carrier signal frequencies with three chipping codes of 10 symbols**

- Therefore the spread spectrum frequency bandwidth is 3.6864 MHz. The bandwidth of a CDMA-2000 BTS is 5 MHz for *n* = 12, 12 carrier channels are transmitted with a time space shift of (1/12) × 814 ns (phase angle change of 30°). Therefore multi-carrier transmission in 3G CDMA-2000 uses multiplexing in time space.
- All channel carriers are chipped using the same orthogonal code and scrambled using the same PN long codes, but the chipping instants of each carrier channel are shifted by (1/*n*) × 814 *ns* (phase angle change of 360°/*n*) in 3G CDMA-2000.

### 5.18.7 Features of MS and BTS in CDMA-2000

- The key features of 3G CDMA-2000 Mobile Station (MS) and Base Transceiver Station (BTS) processing units are as follows :

1. Synchronous base stations.
2. Interleaving of signals of low data rates (between 0.8 kbps and 4.8 kbps).
3. Multi-rate transmission by variable spread factor (between 4 and 256) for data rates of 307.2 kbps to 4.8 kbps.
4. Multi-Carrier (MC) FDD mode-2 transmission by using chipping rate = *n* × 1.2288 Mchip/s, where *n* = 1, 3, 6, 9 or 12 and all *n* carriers have unified power control.
5. Employing distinct uplink and downlink multiple access, modulation, spreading and channelization codes.

### 5.18.8 Advantages

1. It is a very efficient and robust technology.
2. It delivers the highest voice capacity and data throughput using the least amount of spectrum.
3. It can be used to provide services in urban as well as remote areas cost effectively.
4. It supports both voice and data services on the same carrier and allows operators to provide both services cost effectively.
5. It is capable to support large volumes of data traffic at broadband speeds.
6. It is well suited to provide high speed data services to its mobile subscribers and/or broadband access to the Internet.
7. It enables operators to invest in fewer cell sites and deploy them faster ultimately allowing the service providers to increase their revenues with faster Return On Investment (ROI).
8. It can provide operators a significant time to market advantage over other 3G technologies.

### 5.18.9 Advantages

1. Increased voice capacity.
2. Higher data throughput.
3. Multicast services.
4. Frequency band flexibility.
5. Migration paths.
6. Serves multiple market.
7. Supports multiple service platforms.
8. Full backward compatibility.
9. Increased battery life.
10. Synchronization.
11. Power control.
12. Soft hand-off.
13. Transmit diversity.
14. Voice and data channels.
15. Traffic channels.
16. Supplemental channels.
17. Turbo coding.
18. Full backward compatibility.
19. Connectivity to ANSI-41, GSM-MAD and IP networks.
20. Improved service multiplexing and QoS management.
21. Flexible channel structure in support of multiple services with various QoS and variable transmission rates.

## 5.19 IS-95 STANDARDS

### 5.19.1 Brief History

- One of the targets set by the *Cellular Telecommunication Industry Association* (CTIA) for the digital cellular systems for North America was *ten fold* increase in traffic capacity over the analog AMPS system. The capacity increase provided by the TDMA-based D-AMPS, which currently supports three traffic channels per carrier, is only three times the AMPS capacity.
- In the year 1990, Qualcomm, USA developed and demonstrated a CDMA-based digital cellular system that claimed a *twenty fold* increase in capacity over the analog system. In the year 1992, following a request from the CTIA, *Telecommunications Industry Association* (TIA) of USA TR45 initiated standardization work on wide band spread spectrum technologies for cellular applications.
- The IS-95 standard for the CDMA common air interface was adopted in 1993, followed by an enhanced and revised version (IS-95A) in the year 1995. IS-95 was accepted as a CDMA digital cellular standard in the year 1997.

### 5.19.2 Introduction

- IS-95 was designed to be compatible with the existing US analog cellular system (AMPS) frequency band. Hence mobiles and base stations can be economically produced for dual mode operation. It uses 824 to 849 MHz and 869 to 894 MHz with multiple analog channels forming one digital carrier. The standard uses Frequency Division Duplex (FDD) for forward and reverse link, as in the case of GSM systems.
- IS-95 allows each user within a cell to use the same radio channel and users in adjacent cells also use the same radio channel, since this is a direct sequence spread spectrum CDMA system. CDMA completely eliminates the need for frequency planning within a market.

### 5.19.3 Evolution

- The CDMA cellular communication system was created to meet the objectives of the cellular telecommunications industry association (CTIA) user performance requirement (UPR) specification. Some of the UPR objectives include :
  1. increased subscriber capacity of 10 X compared to analog systems,
  2. data communication capabilities, or
  3. advanced information services.
- At the time CDMA cellular communication specification was being developed the IS-54 TDMA cellular communication specification was being created. While this specification met many of the requirements of the CTIA UPR, it did not meet all of the objectives.

- The company Qualcomm created the initial CDMA/cellular communication specification. This proprietary specification was submitted to the telecommunication industry association (TIA) standards development committee and with a few changes, the TIA created the interim standard - 95 (IS - 95) CDMA cellular communication specification.
- The CDMA cellular communication system 'cdma one' defines a complete wireless mobile and fixed wireless communication system and the industry specifications that are required to ensure interoperability for products and services.
- The IS-95 cellular communication specification has evolved through three revisions. These revisions provided for improved voice quality, increased data transmission rates, and offered advanced information services.

## 5.19.4 Definition

- Interium Standard 95 (IS 95) is the first CDMA based digital cellular standard developed by Qualcomm (U.S.A.) in the year 1990 and was accepted as a standard in 1997. The brand name for IS 95 is CDMA one.
- **Qualcomm (YAS DAQ) :** Qcom is an American global telecommunication corporation that designs, manufactures and markets digital wireless telecommunication products and devices based on its code division multiple access (CDMA) technology and other technologies headquartered in San Diego CA (U.S.A.).
- The company operates through four segments :

1. Qualcomm. CDMA technologies (QCT)
2. Qualcomm. Technology Licensing (QTL)
3. Qualcomm. Wireless and Internet (QWI)
4. Qualcomm. Strategic Initiation (QSI)

- IS-95 is a 2G mobile telecommunication standard that uses CDMA, a multiple access scheme for digital radio to send voice data and signaling data (such as a dialed telephone number) between mobile telephones and all sites.

## 5.19.5 Service Aspects

**1. Types of Services :**

- The various types of services that are offered by the IS-95 standard are as under :

(i) Short message service (SMS).
(ii) Slotted paging.
(iii) Over the air activation.
(iv) Enhanced mobile station identities.
(v) Temporary mobile station identities.
(vi) Asynchronous data group 3 Fax.
(vii) Synchronous data.
(viii) Packet data.
(ix) Supplementary services.

**2. Description :**

- The various services that are offered by IS-95 standards are discussed as under :

**(i) Short message service (SMS) :**

- The SMS will support both mobile terminated and mobile originated short messages of upto 255 octets on either the control channel or traffic channel.
- The SMS can also be used to support such applications as telemetry and digital paging.

**(ii) Slotted paging :**

- The slotted paging enables mobiles to wake up for the one to two time slots (80 ms) per slotted paging cycle to listen to incoming pages.
- The factor is intended to conserve mobile battery power.

**(iii) Over the air activation (OTA) :**

- The OTA allows the mobile to be activated by the service provider without third-party intervention.
- It also provides potential for future remote reprogramming of mobile terminals and for software download capabilities.

**(iv) Enhanced mobile station identities (EMSI) :**

- The EMSI provides for use of International Mobile Station Identities (IMSI) based on the ITTS standard (E.212), thereby facilitating international roaming and separation of mobile Directory Numbers (DNs) and mobile terminal identities.

**(v) Temporary mobile station identities (TMSI) :**

- The TMSI allows for the allocation of TMSI by serving Visitor Location Registers (VLRs), which are used on the air interface to maintain user confidentiality and to reduce overall signaling traffic load across the radio interface.

**(vi) Supplementary services :**

- The supplementary services such as call waiting, call forwarding, calling line, ID and similar services that are currently supported by D-AMPS and GSM.

## 5.19.6 Key Features

1. **Key Features :**

The various key features of IS-95 standards are as under :

(i) It has *time diversity* provided by symbol interleaving, error detection and correction coding.

(ii) It has *frequency diversity* provided by the wide band signal of 1.25 MHz.

(iii) It has *space (path) diversity* provided by the dual cell site receive antenna, multipath take receivers and multiple cell sites (soft hand-off).

(iv) The RF power in the cellular system is controlled for its effective working.

(v) The soft hand-off sequence in a CDMA system involves transition from the donar cell to both the donar and the receiving cell and finally to the receiving cell.

(vi) It has higher system capacity in terms of number of calls.

(vii) The CDMA radios based on the spread spectrum are designed to tolerate some level of interference with their overall capacity limited by how well this mutual interference call be controlled.

(viii) It does not need a SIM card.

2. **Description :**

Some of the features of IS-95 CDMA systems are discussed as under :

**(A) Diversity :** The cellular systems are prone to multipath fading and diversity methods of some are required to mitigate the effects of fading. The types of diversity that are available in a CDMA system includes the following :

(i) Time diversity provided by symbol interleaving, error detection and correction coding.

(ii) Frequency diversity provided by the 1.25 MHz wide band signal.

(iii) Space (path) diversity provided by dual cell-site receive antennas, multipath take receivers and multiple cell sites (soft handoff).

**(B) Power Control :** For the CDMA system to work effectively the RF power in the system needs to be controlled. These power requirements have two key components :

(i) All the transmissions from the mobiles must be received at the base stations receiver at approximately the same strength (within 1 dB) even more conditions of the fast multipath fading.

(ii) To maximize the number of users sharing a cell, only the minimum RF power required for reliable commulation should be allowed from the base station transmitter.

**(C) Soft Handoff :** The soft hand-off in a CDMA system results from the system's capability to simultaneously deliver signals to a mobile through more than one cell. Thus the handoff sequence in a CDMA system involves transition from the donar cell to both the donar and the receiving cell and finally to the receiving cell.

**(D) IS-95 CDMA System Capacity :** The key parameters that determine the capacity of a CDMA digital cellular system are as follows :

(i) Processing gain ratio of spreading code information data rate (W/R).

(ii) Ratio of energy per bit to noise power ($E_b/N_o$).

(iii) Voice activity factor.

(iv) Frequency reuse efficiency.

(v) Number of sectors in the call-site antenna.

**(E) Soft Capacity :** The CDMA radios based on the spread spectrum concept are designed to tolerate some level of interference, with their overall capacity limited by how well this mutual interference call be controlled. This capability is especially important when calls might be dropped during handoff because of a lack of available free channels.

## 5.19.7 Frequency Bands

- Frequency bands are the allocation (available use) of radio frequency (RF) spectrum assigned by a regulatory agency for use for specific types of radio communication services. There are two frequency bands designed for IS-95 CDMA cellular communication system (1) band 0, and (2) band 1.
- Band 0 defines the CDMA channels for use in the 800 MHz frequency band and band 1 defines the CDMA channels to use in the 1900 MHz PCs band. It is not possible to use all of the frequencies within the assigned frequency bands due to the need for frequency guard bands in CDMA cellular communication. Frequency band classes assigned for IS-95 CDMA cellular communication system is as shown in Table 5.2.

**Table 5.2 : Frequency bands used in IS-95**

| Band Class | System | Mobile Transmit Frequency Range (MHz) | Base Transmit Frequency Range (MHz) |
|---|---|---|---|
| 0 | North American Cellular | 820 to 849 | 869 to 894 |
| 1 | North American PCS | 1850 to 1919 | 1930 to 1990 |

## 5.19.8 Frequency and Channel Specifications

- The IS-95 is specified for reverse link operation in the 824 to 894 MHz band and 869 to 894 MHz for the forward link. A Personal Communication System (PCS) version of IS-95 has also been designed for international use in the 1800 to 2000 MHz bands. A forward and reverse channel pair is separated by frequency spectrum of 45 MHz for cellular band operation. Many users share a common channel for transmission.
- The maximum user data rate is 9.6 kbps. The IS-95 uses spread spectrum technology. The spreading process is different for the forward and reverse links in the original CDMA specification.
- On the forward link, the user data is encoded using a rate 1/2 convolutional code, interleaved, and spread by one of 64 orthogonal spreading sequences. Each mobile in a given cell is assigned a different spreading sequence, providing perfect separation among the different users. To reduce interference between mobiles that use the same spreading sequence in different cells and to provide the desired wide band special characteristics, all signals in a particular cell are scrambled.
- On the reverse link, a different spreading strategy is used since each received signal arrives at the base station via a different propagation path. The reverse channel user data stream is first convulationally encoded with a rate of 1/3 code. After interleaving, each block of 6 encoded symbols is mapped. Another essential element of the reverse link is tight control of each subscriber's transmitter power, to avoid the *near end-far end* problem that arise from varying received powers of the users. A combination of open-loop and closed-loop power control is used to adjust the transmit power of each in-cell subscriber so that the base station receives each user with the same received power.
- The commands for the closed-loop power control are send at a rate of 800 b/s and these bits are stolen from the speech frames. Without fast power control, the rapid power changes due to fading would degrade the performance of all users in the system.

### 5.19.9 Specifications

The important specifications of IS-95 standard are as under :

1. Number of channels : 20 (798 users per channel)
2. Channel bit rate : 12288 Mb
3. Channel spacing : 1250 kHz
4. Multiple access method : CDMA/FDMA
5. Mobile frequency range : $R_X$ 869 to 894, $T_X$ 824 to 849
6. Duplex method : FDD
7. Modulation : QPSK/0 QPSK
8. Frequencies used : 800 MHz or 1900 MHz.
9. Channel bandwidth : Total 12 MHz with 1.25 MHz for spread spectrum.
10. Data bit rate : 9.6 kb
11. Voice codec : 8 kbps or 13 kbps.
12. SMS service : Upto 120 characters
13. Type of radio interface : CDMA.
14. Type of hand-off : Soft

### 5.19.10 Architecture

**1. Architecture Diagram :**

The architecture of IS-95 CDMA system is as shown in Fig. 5.12.

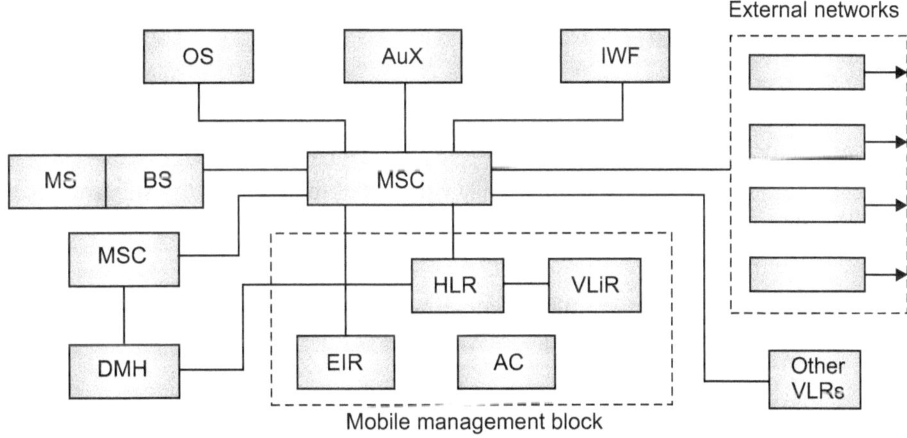


**Fig. 5.12 : Architecture of IS-95 system**

**2. Working :**

- Consider system architecture of IS-95 CDMA system as shown in Fig. 5.12. It consists of the following important elements as under :

  (i) Mobile station (MS).
  (ii) Base station (BS).
  (iii) Mobile switching centre (MSC).
  (iv) Home location register (HLR).
  (v) Data message handler (DMH).
  (vi) Visiting location register (VLR).
  (vii) Authentication centre (AC).
  (viii) Operating system (OS).

- The constituent elements are explained below :

**(i) Mobile station (MS) :**

- The mobile telephone unit is a compact portable handset of the subscriber. It is also known as mobile station (MS). It can be a single unit as standalone or it can be connected to other devices such as fax machine, computer etc.
- It allows the user to have access to the services from the network. It communicates with the base station (BS) over the radio air interface.

**(ii) Base station (BS) :**

- The base station (BS) consists of a base transreceiver station (BTS) and a base station controller (BSC). It is installed for every cell and it communicates with all the mobile stations (MS) existing in the cell.
- It also communicates with the mobile switching centre (MSC). It is responsible for all functions related to the radio channel management. It controls the power level to avoid interference.

**(iii) Mobile switching centre (MSC) :**

- The MSC is the heart of the IS-95 CDMA system. It is a main and centralized switching center which is equivalent to telephone exchange.
- It automatically interfaces the user traffic from the wireless network to the wired network or to the other wireless networks.

**(iv) Home location register (HLR) :**

- The HLR represents a centralized database that has the permanent data fill about the mobile subscribers in a large service area. It is a unit which maintains all the subscriber related information that is required for the management of mobile subscribers.
- It can have a separate existence or it can be an integral part of MSC.

**(v) Data message handler (DMH) :**

- The DMH is assigned the work of collecting the data regarding billing.
- It collects the required data for billing.

**(vi) Visiting location register (VLR) :**

- The VLR represents a temporary data, data store and generally it can be attached to one or more MSCs. It contains the information about the mobile subscribers are currently in the service area and also the information about locally activated features such as call forward on busy.
- It stores the subscriber's information dynamically which is obtained from the subscriber's HLR data. When a roaming MS enters into a new service area, the MSC informs to the corresponding VLR.

**(vii) Authentication center (AC) :**

- The AC can have an independent identity or it can be a part of MSC or HLR. The authentication protects the system network against unauthorized access and the authentication of each subscriber is managed by the AC.
- The AC is responsible for all security aspects and its function is closely linked with the HLR. The purposing of using authentication areas are MS registration, MS organization and MS termination.

**(viii) Equipment identity register (EIR) :**

- The EIR can be located separately or it can be located with the MSC.
- The EIR provides information about the mobile devices which is used for preparing the record.

**(ix) Interworking function (IWF) :**

- The IWF enables the MSC to communicate with other networks.

**(x) External networks :**

- The external networks include other communication networks such as public switched telephone network (PSTN) or integrated service digital network (ISDN), or public switched packet data network (PSPDN) or a public land mobile network (PLMN).

### 5.19.11 Call Processing

1. **Call processing states :**
   - In order to set up a call or transmit data in IS-95 CDMA system, it is necessary to establish a data path through the traffic channel.
   - A mobile station (MS) in IS-95 CDMA system goes through the following four states for the purpose of call processing :

     (i) System initialization state. (ii) System idle state.

     (iii) System access state. (iv) Traffic channel state.

2. **Call processing operation :**

   The operation of call processing in IS-95 CDMA system is discussed as under :

   **(i) System initialization state :**
   - The mobile acquires a pilot channel of a CDMA system. It searches all the PN offset possibilities and selects the strongest pilot signal.
   - It acquires the synchronization channel and detects the pilot channel. It obtains the system configuration and timing information for the CDMA system.

   **(ii) System idle state :**
   - The mobile performs the monitoring procedures of paging channel. It transmits an acknowledgement in response to any message received that is addressed to this mobile.
   - It also maintains all active registration timers.

   **(iii) System access state :**
   - If a cell is being placed or received by the mobile it enters into the access state. In this state it exchanges the necessary parameters. The mobile transmits its response messages or request messages to the base station on the access channel and receives the message from the base station on the paging channel.
   - Similarly the base station transmits its messages to the mobile, the paging channel and receive messages from the mobile on the access channel. The entire process of transmitting one message and receiving an acknowledgement for that message is called an access attempt. The access attempt ends after an acknowledgement is received.

   **(iv) Traffic channel state :**
   - If the access attempt is successful, then the mobile enters into the last state called as traffic state in which the transactions of voice and data take place. The mobile station communicates with the base station using forward and reverse traffic channels, and performs the following functions :

   **(a)** The mobile verifies that it can receive the forward traffic channel and begins transmitting on the reverse traffic channel.

   **(b)** The mobile waits for an order on an alert with the information message.

   **(c)** The mobile waits for the user to answer the call.

   **(d)** The primary service of mobile option application exchanges primary traffic packets with the base station.

   **(e)** The mobile disconnects the call.

### 5.19.12 Radio Aspects

- The spread spectrum techniques used in CDMA cellular systems are adaptations of similar techniques used extensively in military applications since 1950.
- In these techniques, the set of normal payload data is modulated and the desired signal is received by dispreading code.
- At the receiving end the desired signal is received by dispreading the signal with an exact copy of the spreading code in the receiver correlator. Other signals (within the same frequency band) remain fully spread and are perceived as noises.

- The TDMA - based digital cellular systems like GSM and D-AMPS are examples of band-limited systems that aim to maximize the transmitted information rate within the allocated bandwidth by increasing the ratio of bit energy to noise power separate density ($E_bN_o$) for the signal-to-noise ratio (SNR).
- An increase in bandwidth efficiency can be achieved either by selecting a modulation technique like D-OPSK (used in D-AMPS and the Japanese PDC system) that is very agile (i.e. carries an increased number of bits for symbol) or by using a bandwidth conserving modulation technique like GMSK (used in GSM).
- A system based on the spread spectrum concept is an example of a power limited system that is not constrained by bandwidth. Infact, in such systems the transmitted signal is much wider (in frequency spread) than that required for information to be carried.
- Though there are several kinds of spreading techniques, such as direct sequence frequency hopping and chirping the IS-95 CDMA cellular system employs direct sequence spread spectrum method.
- In these systems, the spreading is accomplished by modulating the narrow band information with much wider spreading signal provided by a pseudo-random noise code.
- Since each radio is assigned its own specific PN code, all radios except the desired one appear as noise at the receiver when the composite received signal is correlated with the PN code of the desired radio.
- The PN spreading code is often referred as the chipping code and the resulting bandwidth (after spreading) as the chip rate.
- In the IS-95 CDMA cellular systems, the chip rate is about 1.23 MHz approximately one-tenth of the total 12.5 MHz spectrum (for each direction) allocated to each cellular operator in the United States.
- Thus a set of ten 1.25 MHz bandwidth CDMA channels can be used by each operator, if the entire allocation is converted to CDMA.
- However, in the dual mode CDMA/AMPS environment that is likely to exist for the forescable future, initially only one or a few 1.25 MHz channels need to be assigned to CDMA digital service from the present AMPS analog service.
- Gradually as the demands for digital service increases, more and more analog capacity (increments of 1.25 MHz) can be transferred to CDMA digital operation.
- In a mixed CDMA/AMPS environment, some frequency guard band is required (between the CDMA and AMPS frequency allocation) to ensure that maximum CDMA call carrying capacity can be realized.

### 5.19.13 Network Reference Model and Security Aspects

- The IS-41 standard for inter-network operation has also been enhanced to support the network and signaling requirements for CDMA cellular systems.
- IS-95 CDMA systems also use the authentication and privacy procedures specified in IS-41 which are used in D-AMPS systems.
- Thus, the network reference model and the security aspects for D-AMPS generally also apply for CDMA systems based on the IS-95 standard.

## 5.20 GPRS STANDARDS

### 5.20.1 Brief History

- GPRS opened in 2000 as a *packet switched data service* embedded to the channel-switched cellular radio network GSM. GPRS extends to reach of the fixed internet by connecting mobile terminals worldwide.
- The (ELL) FAC protocol developed in 1991 to 1993, was the trigger point for starting in 1993 specification of standard GPRS by ETSI SMS. Especially, the CELLIPAC *voice* and *data* functions introduced in a 1993 ETSI workshop contribution anticipate what was later known to be the roots of GPRS.
- In 1993, ETSI workshop contribution is referred in 22 GPRS related US-patents. Successor systems to GSM/GPRS like W-CDMA (UMTS) and LTE rely on key GPRS functions for mobile internet access as introduced by CELLIPAC.
- According to the study in the history GPRS development, *Bernhard Walke* and his student *Peter Ducker* are the inverters of GPRS, the first system providing worldwide mobile network access.

### 5.20.2 Introduction

- GPRS was established by ETSI in response to the earlier *CDPD* and *i-mode* packet-switched cellular technologies. It is now maintained by the Third Generation Partnership Project (3GPP).
- GPRS is a *best effort* service, implying variable *throughput* and *latency* that depend on the number of other users sharing the service concurrently, or opposite to *circuit switching*, where a certain Quality of Service (QoS) is guaranteed during the connection. In 2G systems, GPRS provides data rate of 56 to 114 kbps.
- 2G cellular technology combined with GPRS is sometimes described as 2.5 G, that is, a technology between the 2G and 3G of mobile telephony. It provides moderate speed data transfer, by using unused TDMA channels in for example, the GSM system. GPRS is integrated into GSM Release 97 and newer Releases.
- The GPRS core network allows 2G, 3G and *W-CDMA mobile network to transmit* IP packets external networks, such as the internet. The GPRS system is an integrated part of the GSM *network switching subsystem.*

### 5.20.3 Definition

- *The acronym GPRS stands for General Purpose Radio Service. GPRS is a packet oriented mobile data standard on the 2G and 3G cellular communication network global system for mobile communication (GSM).*
- *GPRS is a 3G step toward internet access, which is also known as GSM IP that is a Global System.* Mobile Communications Internet Protocol as it keeps the users of this system online, allows to make voice calls and access internet on the go.

### 5.20.4 Key Features

1. **Always on-line :** It removes the dial-up process, making applications only one click away.
2. **Upgrade to existing systems :** Operators to do not have to replace their equipment, rather, GPRS is added on top of the existing infrastructure.
3. **Integral part of future 3G systems :** GPRS is the packet data core network for 3G systems EDGE and W-CDMA.

### 5.20.5 Protocols Supported

1. Internet protocol (IP)
2. Point-to-point Protocol (PPP)
3. X-25 connections.

### 5.20.6 Goals

1. Open architecture
2. Consistent IP services
3. Same infrastructure for different air interfaces
4. Leverage industry investment in IP
5. Service innovation independent of infrastructure.

### 5.20.7 Architecture

- GPRS architecture works on the same procedure like GSM network, but has additional entities that allow packet data transmission. This data network overlaps a 2G GSM network providing packed data transport at the rates from 3.6 to 171 kbps.
- Along with the packet data transport, the GSM network accommodates multiple users to share the same air interface resources concurrently.
- GPRS attempts to reuse the existing GSM network elements as much as possible, but to effectively build a packed-based mobile cellular networks, same new network elements, interfaces, and protocols for handling packet traffic are required. Therefore, GPRS requires modifications to numerous GSM network elements.

### 5.20.8 Advantages

1. Higher data rate
2. Easy billing
3. Mobility
4. Immediacy
5. Localization

### 5.20.9 Applications

1. Textual and visual information
2. Still images
3. Web browsing
4. Document sharing
5. Corporate E-mail
6. Internet E-mail
7. Vehicle positioning
8. File transfer
9. Fax
10. Intranet/Internet access
11. Value added services (VAS)
12. E-commerce
13. Freight delivery
14. Navigation

## 5.21 GPRS EDGE

### 5.21.1 Introduction

- The next advance in GSM radio access technology was EDGE or E-GPRS. With a new modulation technique yielding a 3 fold increase in bit rate (8 PSK replacing GMSK) and new channel coding for spectral efficiency. EDGE was successfully introduced without disrupting in frequency reuse plans or existing GSM deployment.
- The increase in data speed to 384 kbps placed EDGE as an early pre-state of 3G, although it was labelled as 2.75 G by industry watchers.
- The theoretical maximum speed is 473 kbps for time-slots, but it is typically limited to 135 kbps in order to conserve spectrum resources. Both phone and network must support EDGE, otherwise the phone will travel automatically to GPRS.
- Ongoing standard work on 3 GPP has delivered EDGE evolution, which is designed to complement High-Speed Packet Access (HSPA) coverage. EDGE has improved spectral efficiency with reduced latencies down to 100 ms and increased throughput speeds to 1.3 Mbps in the downlink and 653 Mbps in the uplink. GPRS (Release 97) and EDGE (Release 98) are larger maintained in the RANG Working Group of 3GPP, which succeeded TSA GERAN, when it was closed in 2016.
- EDGE was deployed on GSM networks beginning in2005, initially by **singular** (now AT&T) in the United States. It is considered a pre-3G radio technology and is part of TU's 3G definition. *EDGE is a digital mobile phone technology that allows improved data transmission rates as a backward compatible extension of GSM.*
- EDGE is also standardized by 3 GPP as part of the GSM family. A variant, so called compact. EDGE was developed for use in a portion of Digital AMPs network spectrum.
- Through the introduction of sophisticated methods of coding and transmitting data, EDGE delivers higher bit-rates per radio channel, resulting in a 3 fold increase in capacity and performance compared with an ordinary GSM/GPRS connection. EDGE can be used for any packet switched applications such as in internet connection. Evolved EDGE continues in release 7 of the SGPP standard providing reduced latency e.g., complement HSPA. Peak bit rates of upto 1 mpbs and typical bit-rates of 400 kbps can be expected.

### 5.21.2 Definition

- An acronym of EDGE stand for **Enhanced Data for Global Evolution**. It is also known as Enhance GPRS or E-GPRS.
- *EDGE is a data system used on top of GSM networks, which provides nearly 3 times faster speeds than the outdated GPRS system.*
- EDGE meets the requirements for a 3G network, but it is usually classified as 2.75.
- *GPRS is an evolution of the GSM standard, and for that reason it is sometimes called GSM 1+ or GMS 2+.*

### 5.21.3 Key Features

1. GPRS multi-slot class
2. EDGE multi-slot 12
3. Longer functional life

## 5.22 FORWARD CDMA CHANNEL

### 5.22.1 Structure

- The structure of CDMA channels in the forward link code channels that can be accommodated on each RF channel. The CDMA channels in the forward link are arranged in the following fashion :
  1. Pilot channel
  2. Paging channel
  3. Sync channel
  4. Traffic channel.
- The CDMA channels are associated with different Walsh codes; particular codes being used to support different functions as outlined in Fig. 5.13.
- To generate the final signal, the data from the individual channels is multiplied with the Walsh codes to provide the individual CDMA forward link channels. The output from this process is then further multiplied with the short PN.

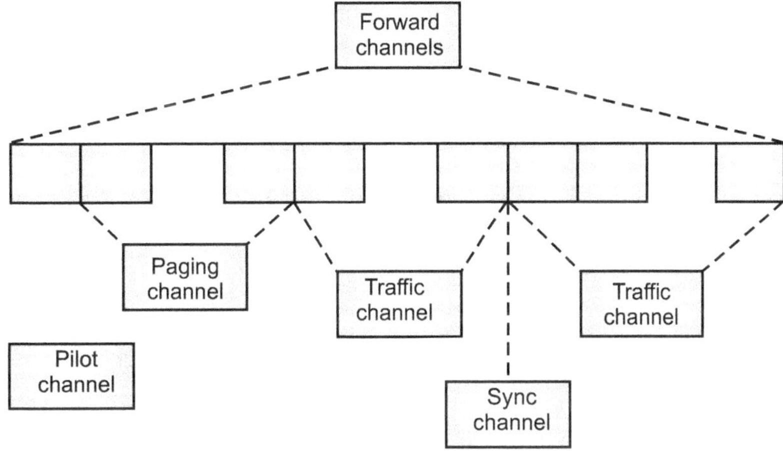


**Fig. 5.13 : CDMA forward channels with Walsh codes**

- To generate the final signal, the data from the individual channels is multiplied with the Walsh codes to provide the individual CDMA forward link channels. The output from this process is then further multiplied with the short PN code. This provides a means of identifying the sector/call from which the signal is coming for the handset/mobile.

### 5.22.2 Classification

1. Pilot channel
2. Paging channel
3. Sync channel
4. Traffic channel

### 5.22.3 Pilot Channel (PC)

- The pilot channel is transmitted as a reference by the base station to provide timing and phase reference for the mobiles, and carries no real data. The *data* carried by the channel is a continuous stream of *zeroes* which is spread by *Walsh code zero,* which is itself a stream of zeroes.
- This is further spread by a pair of quadrature PN sequence with its associated offset. A measurement of the signal-to-noise ratio (SNR) of the pilot channel also gives the mobile an indication which is the strongest service sector.

### 5.22.4 Paging Channel (PC)

- The paging channel is used to carry information to enable mobiles to be pages. Data carried by this channel includes system parameters, voice pages, SMS and other broadcast messages. It occupies Walsh codes 1 to 7 dependent upon the system requirements. It carries data at either 4.8 or 9.6 kbps, a field in the synchronization channel indicates the data rate being transmitted.
- As with other channels, there are a number of stages taken to produce final channel. First the baseband information is error protected. After this, the data is repeated, if it is at a rate of 4.8 kbps, otherwise it is left as it is. Following the data is inter-leaved and then scrambled by the decimated long PN sequence, and finally spread by the Walsh code for the particular assignment.
- In this process, the long PN code is itself masked with a code for paging channel 1 fusing Walsh code is different to paging channel 4 using Walsh code 4.

### 5.22.5 Synchronization Channel (SC)

- Synchronization channel is used to provide the timing reference to access the call. This channel always uses Walsh code 32. Each base station has a fixed timing offset to reduce the interference between adjacent base stations.
- The synchronization channel incorporates an 80 ms superframe, structure. This is divided into three 26.667 ms frames, which corresponds to the same length as the short PN sequences. This means that they align with the timing on the pilot channel.
- This channel is allocated the least power of the overhead channels in the overall CDMA transmission. The data that is transmitted on this channel includes the system times, pilot PN of the base station, long code state, system IP, and the network IP.

### 5.22.6 Forward Traffic Channel (FTC)

- As the name implies, the forward traffic channel (FTC) is used to carry voice, user data, and signaling information. When carrying voice, the coded voice data does not require a constant bit rate and CDMA allows the rate of the frames to change dynamically every 20 ms. When this rate is reduced, it reduces the level of interference to other users.
- The original vacoder specification used a set of rates based on division of 3.6 kbps. This is reflected in CDMA. Later the vacoder was approved to give better voice quality and is a CDMA vacoder was introduced with a rate set on 14.4 kbps was termed RS2. However, data is always carried at full rate.

## 5.23 CDMA DATA CHANNELS

### 5.23.1 Introduction

- As with cellular to telecommunications system, within CDMA different channels are used to carry different forms of data. The CDMA channels can be split into those for the *forward link* and therefore *reverse link.* The CDMA channels differ between the forward and reverse links as a result of different requirements and the different way in which the links operate.
- Also Walsh codes need to be synchronized, if they are to remain orthogonal. As the signals transmitted from the mobile stations travel over different distances, because of the variety of locations of mobiles, they will arrive at slightly different times, and hence they will not be synchronized to one another. The addition of the PN code overcomes this problem, even though Walsh codes have been used for part of the spreading process in the mobile.

### 5.23.2 Classification

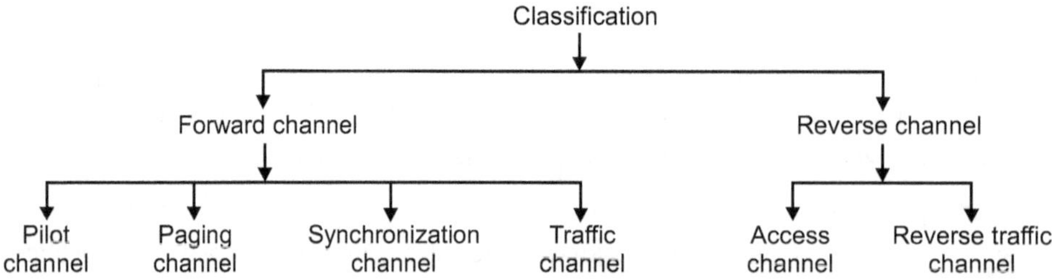


**Fig. 5.14 : Classification of CDMA data channels**

## 5.24 REVERSE CDMA CHANNELS

### 5.24.1 Structure

- The CDMA channel for the reverse link are quite different to those in the forward link. There are only two basic channels, namely **access channel** and **reverse traffic channel.** The way in which these CDMA channels are structured and assembled is also different. This is because they are generated within the mobile rather than the base station.
- In terms of the modulation, OQPSK is used where a half chip delay is introduced onto the Q channel of the modulation. However, orthogonal modulation schemes are used. The different mobiles are individually identified by a mask on the long PN code, which is based on the Equipment Serial Number (ESN). The long PN code is used to give the find spreading of the data to 1.228 Mcps.

### 5.24.2 Forward Channel Modulation

- The forward data appearing on the various CDMA channels is brought together so that it can be summed and modulated onto the radio frequency (RF) carrier. The individual gains of each CDMA channel are adjusted to provide the correct relative teaser for each channel. This is undertaken, because the levels of traffic on each channel are always changing.
- Once the channel gains have been set the signals are added coherently to form the spectrum speed. Then the I and Q components are each modulated onto a carrier and summed to give the final transmitted QPSK signal.

### 5.24.3 Classification

- The classification of CDMA reverse channel are as under :
  1. Access channel.
  2. Reverse traffic channel.

### 5.24.4 Access Channel (AC)

- The CDMA channel to used by the mobile to communicate with the base station, when no traffic channel has been set up. This channel is, therefore, used for gaining occurs to the network, call origination requests and also for sending responses to paging commands and right be sent by the network.
- There can be upto 32 access channels on the CDMA reverse link for each paging channel or forward link. Such access channel uses the same PN, but they are time shifted to enable the mobile to be uniquely identified. Data is sent at 4.3 kbps in a 20 ms time frame, so that each frame contains 90 bits.

### 5.24.5 Reverse Traffic Channel (RTC)

- Like the Forward Traffic Channel (FTC), this reverse link CDMA channel is used to carry variable rate voice data, user data and signaling.
- The structure of the reverse traffic channel is similar to that of the access channel. However, this channel also includes a data burst radometer into which the orthogonally modulated data is fed. The data bursts radometer is the technique used to account for the variable rate voice data is accommodated.
- It is not possible to use the same technique used on the forward link, because they affect the channel power. This in turn would upset the power control adjustments that need to be made to ensure that all the mobiles are received as dose to the same strength as possible.
- Although there are only for CDMA channels on the reverse link, there are all that are needed to carry the required amount of data from the mobile to the base station.

## 5.25 WIRELESS CABLE TELEVISION

### 5.25.1 Brief History

- Cable television began in the United States as a commercial business in 1950, although there were small-scale systems by *nobbiest* in the 1940s. The early systems simply received weak *broadcast* channels, amplified them, and sent them over unshielded wires to the subscribers, limited to a community or to adjacent communities.
- The receiving antenna would be higher than any individual subscriber could afford, thus bringing in stronger signals, in hilly or mountainous terrain it would be placed at high elevation.
- At the outset, cable systems only served smaller communities without television stations of their own, and which could not easily receive signals from stations in cities, because of distance or hilly terrain. In Canada, however, communities with their own signals were ferrite cable markets, as viewers wanted to receive *American* signals. Rarely as in the college town of *Alfred, New York,* U.S. cable systems retransmitted *Canadian* channels.
- Although early VHF television receivers could receive 12 channels (2 to 13), the maximum number of channels that could be broadcast in one city was 7 : channels 2, 4, either 5 or 6, 7, 9, 11 and 13, as receivers at that time were unable to receive strong local signals on adjacent channels without distortion. There were frequency gaps between 4 and 5, and between 6 and 7, which allowed both to be used in the same city.
- As equipment improved, all 12 channels could be utilized, except where a local VHF television station broadcast. Local broadcast channels were not usable for signals deemed to be priority, but technology allowed low priority signals to be placed on such channels by synchronizing their *blanking intervals.*
- Similarly, a local VHF station could not be carried on its broadcast channel as the signals would arrive at the TV set slightly separated in time, causing *ghosting*. The bandwidth of the amplifiers also was limited, meaning frequencies over 150 MHz were difficult to transmit to distant portions of the coastable network and UHF channels had to be used, located between the FM band and channel 7, or *superband* beyond channel 13 upto about 300 MHz; these channels initially were only accessible using separate tuner boxes that sent the chosen channel into the TV set on channel 2, 3 or 4.
- Before being added to the cable by itself, these mid-band channels were used for easily incarnations of pay TV. Once, tuners that could receive select mid-band and super-band channels began to be incorporated into standard TV sets, broadcasters were forced to either install scrambling circuitry or more these signals further out of the range of reception for early cable-ready TVs and VCRs. However, once consumer sets had the ability to receive 11 FCC allocated channels, premium broadcasters were left with not choice put to scramble.
- Unfortunately for pay TV operators, the descrambling circuitry was often published in electronics hobby magazines such as *Bipolar Science* and *Popular Electronics*, allowing anybody with anything more than a knowledge of broadcast knowledge electronics to be able to built their own and receive the programming without cost.
- Later, cable operator began to carry FM radio stations, and encouraged subscribers to connect their FM stereo sets to cable. Before stereos and *bilingual TV sound* became common, Pay-TV channel sound was added to the FM stores cable line-ups.
- About this time, operators expanded by the 12 channels dial to use the *midband* and *superband* VHF channels adjacent to the *'highband'* 7 to 13 of *North American television frequencies.* Some operators in *cornwall, Ontario,* used a dual dists button, network with channels 2 to 13 on each of the two cables.
- During the 1980s, United States requisitions not unlike *Public, Educational* and *Government access* (PEG) created the beginning of cable originated *live television programming.* As cable penetration increased, numerous cable only TV stations were launched many with their own news beiraus localized content than that provided by the nearest network newscast.

- Such stations may use similar on-air branding as that used by the nearly broadcast network affiliate, but the fact these stations do not broadcast over the air and are not regulated by the FCC, their call signs are meaningless. These stations evolved, partially into today's over the air digital sub channels, where a main broadcast TV stations, many live local programs with local interests were subsequently created all over the United States in most major television markets in the early 1980s.
- This evolved into today's many cable only broadcasts of diverse programming, including cable only produced *television movies* and *nano casting*. Cable *specially channels,* starting with channels oriented to show moves and large sporting or performance events classified further, and *narrow casting* because common.
- By the late 1980s, cable-only signals outnum based, broadcast signals on cable systems, some of which by this time had expanded beyond 35 channels the mid-1980s in Canada, cable operators were allowed by the regulations the enter into distribution contracts with cable networks in their own.
- By the 1990s, tiers became common with customers able to subscriber to different tiers to obtain different selections of additional channels above the basic selection. By subscribing to additional tiers, customers could get speciality channels, movies channels and foreign channels.
- Large cable companies used addressable discrambles to limit access to *premium channels* for customers not subscribing to higher tiers, however, the above magazines often published work arounds for that technology as well.
- During the 1990s, the pressure to accommodate the granting array of offering resulted in digital transmission that made more efficient use of the VHF signal capacity. Fibre optics was common to carry signals into areas near the home, where coaxial cables could only higher frequencies over the short remaining distance.
- Although, for a time in the 1980s and 1990s, the television receivers and VCRs were equipped to receive the *mid-band* and *super-band* channels. Due to the facts that the descrambling circuitry was for a time present in these tuners, depriving the cable operator of much of their revenue, such a return to the *set-top boxes* used from the 1970s onward.
- The conversion to digital broadcasting has put all signals broadcast and cable into digital form, rendering analog cable television service mostly obsolete, functional in an error or dwindling supply of select markets. Analog television sets are still accommodated, but their tuners are mostly obsolete, often times dependent entirely on the set-top box.
- Cable television is mostly available North America, Europe, Australia, South Asia and East Asia and lesser in South American and the Middle East. Cable television has had title success in Africa, as it is not cost-effective to lay cables in sparsely populated areas. So called *'wireless cable'* or microwave based systems are used instead.

## 5.25.2 Definition

- *Cable television is a system of delivering television programming to consumers via radio frequency (RF) signals transmitted through coaxial cables, or in more recent systems, light pulses through fiber optic cables.*

## 5.25.3 Key Features

1. Infrared (IR) remote control.
2. RF channel synchronization.
3. Infrared (IR extender.
4. On-screen menue.
5. Built-in analog cable TV tuner.
6. Watching digital channels.
7. RCA/composite output on RX
8. Input and output on TV.
9. RF coaxial splitter.
10. Two-way communication.
11. 4 selectable frequencies.

### 5.25.4 Wireless Cable

- Wireless cable is a name given to a service that is called Multichannel Multipoint Distribution Service (MMDS). It is a type of cable television (TV) system that is its subscribers mix of satellite channels by transmitting the programming over MMDS frequencies along with MMD, OFS, and IIs frequencies, if they are available.
- Wireless cable uses Super High Frequency (SHF) channels to transmit satellite cable programming over the air instead of through overhead or underground wires.
- This contrast with *broadcast television*, in which the television signal is transmitted over the air by radio waves and received by a *television antenna* attached to the television; or a *satellite television*, in which the television signal is transmitted by *communication satellite* orbiting the Earth and received by a *satellite dish*.
- FM radio programming, high speed internet, telephone services and similar non-television services may also be provided through these cables. Analog television was a standard in the 20$^{th}$ century, but since the 2000s, cable systems have been upgraded to digital cable operation.
- A *cable channel* sometimes known as a *cable network* is a television network available via cable televisions. When available through satellite television, including direct broadcast satellite providers.
- Alternative terms include *'non-broadcast channel'* or *'programming services'*, the later being mainly used in legal contexts.
- The abbreviation CATV is often used for *Cable Television.* It originally stood for *Community Access Television* or *Community Antenna Television,* from cable televisions origin in 1948. In areas where over theair TV reception was limited by distance from transmitters or mountains terrain, larger *community antennas* were constructed and cable was run from them to individual homes.

### 5.25.5 Principle of Operation

- In the most common systems, multiple television channels are distribute the subscriber residences through a *coaxial cable,* which comes from a trunklink supported on *utility poles* originating at the cable company's local distribution facility, called the *'headend'*. Many channels can be transmitted through one coaxial cable by a technique called FDMA. At the headend, each television channel is translated to different *frequency 'slot'* on the cable, the separate television signals do not interfere with each other.
- At an *outdoor cable box* on the subscribers residence, the company's service drop cable is connected to cables distributing the signal to different rooms in the building. At each television, the subscriber's television or set-top box provided by the cables company translates the desired channel back to its original frequency (*baseband*) and it is displayed on the screen.
- Due to wide spread *cable theft* in earlier analog systems, the signals are typically *encrypted* in modern digital cable systems, and the set-top box must be activated by an *activation code* sent by the cable company before it will function, which is only sent after the subscriber sign up.
- If the subscriber fails to pay their bill, the cable company can send a signal deactivate the subscriber's set-top box, preventing reception. There are usually *'upstream'* channels on the cable to send data from the customer set-top box to the cable head end, for advanced features, such as requesting *pay per view* shows or movies, *cable internet access* and *cable telephone services.*
- The *'down stream'* channels occupy a band of frequencies from approximately 50 MHz to 1 GHz, while the *'up stream'* channels occupy frequencies of 5 to 42 MHz. Subscribers pay with a monthly for subscribers can choose from several levels of service with *'premium'* packages including more channels, but costing at higher rates.
- At the local *headend,* the fixed signals from the individual television channels are received by dish antennas from communication satellites. Additional local channels, such as local *broadcast television* stations, *educational channels* from local colleges, and *community access* channels devoted to local government (PEG) channels are usually included on the cable service. Commercial advertisements for local business are also inserted in the programming at the headend (individual channels) which are distributed nationally, also have their own nationally oriented commercials.

### 5.25.6 Benefits/Advantages

1. Watch different channels on two different TVs.
2. Save extra cable box and monthly fees.
3. No wires to run and no holes to drill.
4. RF channel synchronization.
5. Extra receivers available to transmit to multiple TVs.
6. Infrared (IR) remote extender.
7. Eliminates interference.

## Practice Questions

1. What is wireless communication ?
2. State advantages of wireless communication.
3. List applications of wireless communications.
4. What is a wireless system ?
5. List types of mobile radio system standards.
6. Write short notes on the following signals :
   (a) AMPS
   (b) ETACS
   (c) IS-54
   (d) IS-136
   (e) GSM.
7. State key features of GSM.
8. What is GSM ?
9. List characteristics of GSM.
10. Describe important GSM services.
11. Draw architecture of GSM and explain its working principle.
12. Describe subsystems of GSM.
13. Draw block diagram of a GSM and describe its each block.
14. Give classification of GSM channels.
15. What are physical and logical channels ?
16. Show GSM frame structure.
17. What is speech processing ?
18. Which is speech ? What is speech processing ?
19. Show speech structure.
20. State important applications of speech processing.
21. What is 3G technology ?
22. List key features of 3G technology.
23. State advantages and disadvantages of 3G technology.
24. State important applications of 3G technology.
25. What is UMTS standard ? State air-interface structure.

26. Draw 3G UMTS network architecture.
27. Explain UMTS security procedure.
28. What is CD-2000 standard ?
29. State key features of CD-2000.
30. Describe CD-2000 standard.
31. State advantages of CD-2000 standard.
32. List important applications of CD-2000.
33. What is IS-95 standard ?
34. List various services offered by IS-95.
35. State key features of IS-95.
36. List specifications of IS-95.
37. Draw architectural diagram of IS-95 and explain its working.
38. What is GPRS ? State key features of GPRS.
39. State advantages of GPRS.
40. State important applications of GPRS.
41. What is GPRS EDGE ?
42. State key features of GPRS EDGE.
43. State applications of GPRS EDGE.
44. List advantages of GPRS EDGE.
45. What is forward CDMA channel ?
46. Describe the following :
    (a) Pilot channel
    (b) Paging channel
    (c) Synchronous channel
    (d) Traffic channel
47. What is CDMA data channel ? Show the structure of forward channel.
48. What is reverse CDMA channel ?
49. Describe access channel and reverse traffic channel.
50. What is wireless cable TV ?
51. Explain principle of operation of wireless cable TV.

✍✍✍

# FUTURE TRENDS

**Syllabus**

4G, 5G and Other Higher G Mobile Techniques, LTE-Advance Systems.

## 6.1 4G TECHNOLOGY

### 6.1.1 Brief History

- The 4G system was originally envisioned by the DARPA (US **D**efense **A**dvanced **R**esearch **P**roject **A**gency). It selected the distributed architecture and end-to-end Internet Protocol (IP), and believed at an early stage in *peer-to-peer networking*, in which every mobile device would be both a transceiver and a router for other devices in the network, eliminating the spokre-and-hub weakners 2G and 3G cellular systems.
- Since, the 2.5 G GPRS system, cellular systems have provided dual infrastructures, *pocket switched nodes* for data services, and *circuit switched network nodes* for voice calls. In 4G systems, the circuit-switched infrastructure is provided, while 2.5 G and 3G systems require both packet-switched and circuit *network nodes,* i.e. two infrastructures in parallel. This means that in 4G traditional voice calls are replaced by *IP telephony*.
- In 2002, the strategic vision for 4G, which ITU designated as IMT Advanced, was laid out. In 2004, LTE was first proposed by NTT **DoCoMo** of Japan. In 2005, OFDMA transmission technology is chosen as candidate for the HSOPA downlink, later renamed 3 GRP was Long Term Evaluation (LTE) as air interface E-UTRA. In November 2005, **KT corporation** demonstrated mobile Wi-MAX service in Busan, South Korea.
- In April 2006, **KT corporation** started the world's first commercial mobile WiMAX service in Seoul, South Korea. In mid-2006, **Sprint** announced that it would invest about US$5 billion in a WiMAX technology buildout over the next few years. Since that time **Sprint** has faced many setbacks that have resulted in steep quarterly losses.
- On 7 May 2008, *Sprint, Imagine, Google, Intel, Comcast, Bright House,* and *Time Warner* announced a pooling of an average of 120 MHz of spectrum; **Sprint** merged its Xohm WiMAX division with Clearwire to form a company which will take the name **'Clear'**. In February 2007, the Japanese company NTT DoCoMo tested a 4G communication system prototype with $4 \times 4$ MIMO called VSF-OFCDM at 100 Mbit/s while moving, and 1 Gbps, while stationary.
- NTT DoCoMo completed a trial in which they reached a maximum packet transmission rate of approximately 5 Gbps in the downlink with $12 \times 12$ MIMO using a 100 MHz frequency bandwidth while moving at 10 km/h, and is planning on releasing the first commercial network in 2010.
- In September 2007, NTT DoCoMo demonstrated e-UTRA data rates of 200 Mbps with power consumption below 100 mW during the test. In January 2008, a U.S. Federal Communications Commission (FCC) spectrum auction for the 700 MHz former analog TV frequencies began. As a result, the biggest share of the spectrum went to Verizon Wireless and the next biggest to AT&T. Both of these companies have started their intention of supporting LTE.
- In January 2008, EU commissioner Viviane Reding suggested re-allocation of 500-800 MHz spectrum for wireless communication, including WiMAX. On 15 February 2008, Skywer Solutions released a front-end module for e-UTRAN. In November 2008, ITU-R established the detailed performance requirements of IMT-Advanced, by issuing a Circular Letter calling for candidate Radio Access Technologies (RATs) for IMT-Advanced.

- In April 2008, just after receiving the circular letter, the 3GPP organized a workshop on IMT-Advanced where it was decided that LTE Advanced, an evolution of current LTE standard, will meet or even exceed IMT-Advanced requirements following the ITU-R agenda. In April 2008, LG and Nortel demonstrated e-UTRA data rates of 50 Mbit/s while travelling at 110 km/h.
- On 12 November 2008, HTC announced the first WiMAX-enabled mobile phone, the *Max 4G*. On 15 December 2008, *San Miguel Corporation*, the largest food and beverage conglomerate in Southeast Asia, has signed a memorandum of understanding with Qata Telecom QSC (*Qtel*) to build wireless broadband and mobile communications projects in the Philippines. The joint-venture formed *wi-tribe Philippines*, which offers 4G in the country. Around the same time *Globe Telecom* rolled out the first WiMAX service in the Philippines.
- On 3 March 2009, Lithuania's LRTC announcing the first operational "4G" *mobile WiMAX* network in Baltic states. In December 2009, *Sprint* began advertising "4G" service in selected cities in the United States, despite average download speeds of only 3-6 Mbps with peak speeds of 10 Mbps.
- On 14 December 2009, the first commercial LTE deployment was in the Scandinavian capitals *Stockholm* and *Oslo* by the Swedish-Finnish network operator *TeliaSonera* and its Norwegian brandname *NetCom* (Norway).
- On 4 June 2010, *Sprint* released the first *WiMAX* smartphone in the US, the HTC Evo 4G. On November 4, 2010, the *Samsung* Craft offered *MetroPCS* is the first commercially available LTE smartphone. On 6 December 2010, at the *ITU World Radiocommunication Seminar* 2010, the ITU stated that LTE, *WiMax* and similar "evolved 3G technologies" could be considered "4G".
- In 2011, *Argentina's Claro* launched a pre-4G HSPA+ network in the country. In 2011, Thailand's Truemove-H launched a pre-4G HSPA+ network with nationwide availability. On March 17, 2011, the *HTC Thunderbolt* offered by Verizon in the U.S. was the second LTE smartphone to be sold commercially. In February 2012, *Ericsson* demonstrated mobile-TV over LTE, utilizing the new eMBMS service (enhanced Multimedia Broadcast Multicast Service).
- Since 2009, the LTE-Standard has strongly evolved over the years, resulting in many deployments by various operators across the globe. For an overview of commercial LTE networks and their respective historic development. Among the vast range of deployments many operators are considering the deployment and operation of LTE networks.

### 6.1.2 Introduction

- In the field of mobile communications, a "*generation*" generally refers to a change in the fundamental nature of the service, non-backwards-compatible transmission technology, higher peak bit rates, new frequency bands, wider channel frequency bandwidth in Hertz, and higher capacity for many simultaneous data transfers **(higher system spectral efficiencyin bit/second/Hertz/site)**.
- New mobile generations have appeared about every ten years, since the first move from 1981 *analog* (1G) to *digital* (2G) transmission in 1992. This was followed, in 2001, by 3G multi-media support, *spread spectrum* transmission and, at least, 200 kbit/s peak bit rate, in 2011/2012 to be followed by "real" 4G, which refers to all-*Internet Protocol* (IP) *packet-switched* networks giving mobile ultra-broadband (gigabit speed) access.
- While the ITU has adopted recommendations for technologies that would be used for future global communications, they do not actually perform the standardization or development work themselves, instead relying on the work of other standard bodies such as IEEE, WiMAX Forum, and 3GPP.
- In the mid-1990s, the ITU-R standardization organization released the IMT-2000 requirements as a framework for what standards should be considered 3G systems, requiring 200 kbps peak bit rate. In 2008, ITU-R specified the IMT Advanced (International Mobile Telecommunications Advanced) requirements for 4G systems.
- The fastest 3G-based standard in the UMTS family is the HSPA+ standard, which is commercially available since 2009 and offers 28 Mbps downstream (22 Mbps upstream) without MIMO, i.e. only with one antenna, and in 2011 accelerated upto 42 Mbps peak bit rate downstream using either DC-HSPA+ (simultaneous use of two 5 MHz UMTS carriers) or $2 \times 2$ MIMO.

- In theory, speeds upto 672 Mbps are possible, but have not been deployed yet. The fastest 3G-based standard in the CDMA 2000 family is the EV-DO Rev. B, which is available since 2010 and offers 15.67 Mbps downstream.

### 6.1.3 Definition

- **4G is the fourth generation of broadband cellular network technology, succeeding 3G.**
- A 4G system must provide capabilities defined by ITU in IMT Advanced.
- **A 4G is the fourth generation of mobile phone technology, which follows on from the existing 3G (third generation) and 2G (second generation) mobile technology.**
- 4G technology builds upon what 3G offers, but does everything at a **much faster speed**.
- **A 4G is defined as the fourth generation of mobile technology which follows the 2G and 3G networks that came before it.**
- It is also sometimes referred to as 4G LTE, but this is not technically correct as LTE is only a single type of 4G. It is currently the most advanced technology that is adopted by the majority of mobile network service providers.

### 6.1.4 Importance

- When it initially came out, 4G quickly changed how we use mobile internet. While 3G networks were relatively fast, 4G network connections allowed users to browse the web and stream HD videos on mobile devices, which basically turned smartphones into the computers of the modern age. 4G data networks have revolutionized the mobile industry and paved the way for the widespread use of today's mobile devices.
- Another advantage of 4G networks is its much lower latency. Low latency is important when real-time interaction is required, for instance, in video conferencing or online gaming. Unified Communication apps used on mobile devices remain clear, responsive, and fast – helping distant colleagues collaborate and work together seamlessly as if they were in the same office.
- What is more, 4G was also immensely useful in the Voice over LTE (VoLTE) industry. It provides a much more stable and clearer connection for voice calls, making it easier for businesses to switch to internet-based telephone systems.
- A benefit to enterprise businesses is the ease and speed of 4G network set up. If a company needs to quickly set up a network in a new location, 4G can be installed in a matter of hours, compared to days or even weeks of setting up a wired connection.
- Even the automotive industry has started using LTE-based car connectivity features, some even offering 4G network hotspots in their vehicles. Hotspots have also been shown to be incredibly helpful to remote workers when a WiFi connection isn't feasible.
- The importance and impact of 4G networks stretch through numerous other industries, as companies and individual users now have the freedom to stay connected in a less restrictive way. Mobile networks will become more important as new uses and applications are discovered. Companies of all sizes find that a 4G network solution makes sense. The flexibility, speed, and reliability that it can offer are hard to match by any other type of technology.

### 6.1.5 Requirements

- An IMT-Advanced cellular system must fulfill the following requirements:
  1. Be based on an all-IP packet switched network.
  2. Have peak data rates of upto approximately 100 Mbps for high mobility such as mobile access and upto approximately 1 Gbps for low mobility such as nomadic/local wireless access.
  3. Be able to dynamically share and use the network resources to support more simultaneous users per cell.
  4. Use scalable channel bandwidths of 5 to 20 MHz, optionally upto 40 MHz.

5. Have peak link **spectral efficiency** of 15 bps·Hz in the downlink, and 6.75 bps·Hz in the uplink meaning that 1 Gbps in the downlink should be possible over less than 67 MHz bandwidth.
6. System spectral efficiency is, in indoor cases, 3 bps·Hz·cell for downlink and 2.25 bps·Hz·cell for uplink.
7. Smooth handovers across heterogeneous networks.

### 6.1.6 Key Features

1. **MIMO:** To attain ultra high spectral efficiency by means of spatial processing including multi-antenna and multi-user MIMO
2. **Turbo principle error-correcting codes:** To minimize the required SNR at the reception site.
3. **Channel-dependent scheduling:** To use the time-varying channel.
4. **Link adaptation:** Adaptive modulation and error-correcting codes.
5. **Mobile IP** utilized for mobility.
6. **IP-based femtocells** (home nodes connected to fixed Internet broadband infrastructure).

### 6.1.7 Advantages

1. Improved download/upload speeds.
2. Reduced latency.
3. Crystal clear voice calls.

### 6.1.8 Disadvantages

1. 4G introduces a potential inconvenience for those who travel internationally or wish to switch carriers.
2. In order to make and receive 4G voice calls, the subscriber handset must not only have a matching frequency band (and in some cases require unlocking), it must also have the matching enablement settings for the local carrier and/or country.
3. While a phone purchased from a given carrier can be expected to work with that carrier, making 4G voice calls on another carrier's network (including international roaming) may be impossible without a software update specific to the local carrier and the phone model in question, which may or may not be available (although fallback to 3G for voice calling may still be possible if a 3G network is available with a matching frequency band).

## 6.2 5G TECHNOLOGY

### 6.2.1 Brief History

- In April 2008, NASA partnered with *Geoff Brown* and *Machine-to-Machine Intelligence (M2Mi) Corporation* to develop 5G communications technology. In 2008, the *South Korean* IT R & D program of "*5G mobile communication systems based on beam-division multiple access and relays with group cooperation*" was formed.
- In August 2012, *New York University* founded NYU WIRELESS, a multi-disciplinary academic research centre that has conducted pioneering work in 5G wireless communications.
- On 8 October 2012, the UK's *University of Survey* secured £35M for a new 5G research centre, jointly funded by the British government's UK Research Partnership Investment Fund (UKRPIF) and a consortium of key international mobile operators and infrastructure providers, including Huawei, Samsung, Telefonica Europe, Fujitsu Laboratories Europe, Rohde & Schwarz, and Aircom International.
- It will offer testing facilities to mobile operators keen to develop a mobile standard that uses less energy and less radio spectrum, while delivering speeds faster than current 4G with aspirations for the new technology to be ready within a decade.
- On 1 November 2012, the EU project "*Mobile and wireless communications Enablers for the Twenty-twenty Information Society*" (METIS) starts its activity toward the definition of 5G. METIS achieved an early global consensus on these systems. In this sense, METIS played an important role of building consensus among other external major stakeholders prior to global standardization activities. This was done by initiating and addressing work in relevant global fora (e.g. ITU-R), as well as in national and regional regulatory bodies.

- Also in November 2012, the iJOIN EU project was launched, focusing on "*small cell*" technology, which is of key importance for taking advantage of limited and strategic resources, such as the radio wave spectrum. According to *Günther Oettinger*, the *European Commissioner* for Digital Economy and Society (2014–2019), "*an innovative utilization of spectrum*" is one of the key factors at the heart of 5G success. Oettinger further described it as "*the essential resource for the wireless connectivity of which 5G will be the main driver*". iJOIN was selected by the *European Commission* as one of the pioneering 5G research projects to showcase early results on this technology at the Mobile World Congress 2015 (Barcelona, Spain).
- In February 2013, ITU-R Working Party 5D (WP 5D) started two study items: (1) Study on IMT Vision for 2020 and beyond, and; (2) Study on future technology trends for terrestrial IMT systems. Both aiming at having a better understanding of future technical aspects of mobile communications toward the definition of the next generation mobile.
- On 12 May 2013, *Samsung Electronics* stated that they had developed a "5G" system. The core technology has a maximum speed of tens of Gbps (gigabits per second). In testing, the transfer speeds for the "5G" network sent data at 1.056 Gbps to a distance of up to 2 kilometers with the use of an $8 \times 8$ MIMO.
- In July 2013, India and Israel agreed to work jointly on development of fifth generation (5G) telecom technologies. On 1 October 2013, NTT (Nippon Telegraph and Telephone), the same company to launch world's first 5G network in *Japan*, wins Minister of Internal Affairs and Communications Award at CEATEC for 5G R&D efforts.
- On 6 November 2013, *Huawei* announced plans to invest a minimum of $600 million into R&D for next generation 5G networks capable of speeds 100 times faster than modern LTE networks. On 3 April 2019, *South Korea* became the first country to adopt 5G. Just hours later, Verizon launched its 5G services in the United States, and disputed South Korea's claim of becoming the world's first country with a 5G network, because allegedly, South Korea's 5G service was launched initially for just six South Korean celebrities so that South Korea could claim the title of having the world's first 5G network.
- In fact, the three main South Korean telecommunication companies (SK Telecom, KT, and LG Uplus) added more than 40,000 users to their 5G network on the launch day. In June 2019, Philippines became the first in Southeast Asia to roll out 5G network after Globe Telecom commercially launched its 5G data plans to customers.
- AT & T bring 5G service to consumers and businesses in December 2019 ahead of plans to offer nationwide 5G in the first half of 2020.

### 6.2.2 Introduction

- 5G is the fifth generation wireless technology for digital cellular networks that began wide deployment in 2019. As with previous standards, the covered areas are divided into regions called "*cells*", serviced by individual antennas. Virtually every major telecommunication service provider in the developed world is deploying antennas or intends to deploy them soon. The frequency spectrum of 5G is divided into millimeter waves, mid-band and low-band. Low-band uses a similar frequency range as the predecessor, 4G.
- 5G millimeter wave is the fastest, with actual speeds often being 1–2 Gbps down. Frequencies are above 24 GHz reaching upto 72 GHz which is above the extremely high frequency band's lower boundary. The reach is short, so more cells are required. Millimeter waves have difficulty traversing many walls and windows, so indoor coverage is limited.
- 5G mid-band is the most widely deployed, in over 20 networks. Speeds in a 100 MHz wide band are usually 100–400 Mbps down. In the lab and occasionally in the field, speeds can go over a gigabit per second. Frequencies deployed are from 2.4 GHz to 4.2 GHz. Sprint and China Mobile are using 2.5 GHz, while others are mostly between 3.3 and 4.2 GHz, a range which offers increased reach. Many areas can be covered simply by upgrading existing towers, which lowers the cost.
- 5G low-band offers similar capacity to advanced 4G. In the *United States*, *T-Mobile* and *AT&T* launched low-band services on the first week of December 2019. T-Mobile CTO *Neville Ray* warns that speeds on his

600 MHz 5G may be as low as 25 Mbps down. AT&T, using 850 MHz, will also usually deliver less than 100 Mbps in 2019. The performance will improve, but cannot be significantly greater than robust 4G in the same spectrum.

- *Verizon*, AT&T, and almost all 5G providers in 2019 have latencies between 25–35 milliseconds. The "*air latency*" (between a phone and a tower) in 2019 equipment is 8–12 ms. The latency to the server, farther back in the network, raise the average to ~30 ms, 25 to 40% lower than typical 4G deployed. Adding "*Edge Servers*" close to the towers can bring latency down to 10 to 20 ms. Lower latency, such as the often touted 1 ms, is years away and does not include the time to the server.
- The industry project 3GPP defines any system using "5G NR" (5G New Radio) software as, "5G", [citation needed] a definition that came into general use by late 2018. Previously, some reserved the term for systems that deliver frequencies of 20 GHz shared called for by ITU IMT-2020. 3GPP will submit their 5G NR to the ITU. In addition to traditional mobile operator services, 5G NR also addresses specific requirements for private mobile networks ranging from industrial *IoT* to critical communications.

### 6.2.3 Need

- The world is going mobile and we are consuming more data every year, particularly as the popularity of video and music streaming increases. Existing spectrum bands are becoming congested, leading to breakdowns in service, particularly when lots of people in the same area are trying to access online mobile services at the same time.
- 5G is much better at handling thousands of devices simultaneously, from mobiles to equipment sensors, video cameras to smart street lights. Therefore, there is a need of 5G technology.

### 6.2.4 Definition

- *5G is the fifth generation of mobile internet connection which differs much faster data download and upload speeds.* Though greater use of the radio spectrum, it will allow for more devices to access the mobile internet at the same time.
- 5G is the next step beyond 4G and LTE mobile networks, with faster speeds, more bandwidth and a wider range than previous 4G networks.
- 5G is a developing technology with several major mobile network carriers researching and improving the network as it events.

### 6.2.5 Concept

- **5G simply refers to the next and newest mobile wireless standard based on the IEEE 802.11 accounting standard of broadband technology.**
- We can say that 5G *wireless technology* denotes the proposed next major phase of mobile telecommunication standard beyond and the current 4G standards.
- Rather that faster internet connection speeds, 5G planning aims at a higher capacity than current 4G, allowing a higher number of mobile broadband users per area unit, and allowing consumption of higher or unlimited data quantities in gigabyte 1 GB per minute and uses.
- This would make it feasible for a large portion of the population to consume high-quality streaming media many hours per day and their mobile devices, also when at of reach of wi-fi hotspots.
- 5G research and development (R&D) also aims at the improved support of machine to machine. Communication, also known as the *Internet of things*, aiming at lower cost, lower battery consumption, and lower latency than 4G equipment.
- Although it is 100 early to decide on what exactly 5G wireless technology is.

### 6.2.6 Basic Requirements

1. High and increased peak bit rate (upto 10 Gbps) connection to end point in the field.
2. Efficient use of energy in device.
3. Larger data volume per unit area i.e. high system spectral efficiency.
4. High capacity to allow more devices connectivity concurrently and instantaneously (100% coverage).

5. More bandwidth.
6. Lower battery consumption.
7. Battery connectivity irrespective of the geographic region, in which you are.
8. Larger number of supporting devices (10 to 100% number of commercial devices).
9. Higher reliability of the communications (one milli-second end-to-end round trip delay).
10. Lower cost of infrastructural development.

- With a large areas of innovation features of 5G, now your smartphone would be more parallel to the laptop.

## 6.2.7 Key Features

1. Network based on the user experience.
2. Enhanced system performance.
3. Business models, management and operations.
4. Beam Division Multiple Access (BDMA) technology.
5. Filter Bank Multicarrier (FBMC) multiple access.
6. For computing for achieving low latency, high mobility, high scalability and real-time executing.
7. Ultra Wide Band (UWB) networks.
8. World combination service mode (WSGM).

## 6.2.8 Working Principle

- Unlike LTE, 5G operates on three different spectrum bands. While this may not seem important, it will have a dramatic effect on your everyday use.
- **Low-band spectrum** can also be described as sub 1GHz spectrum. It is the primary band used by carriers in the U.S. for LTE, and bandwidth is nearly depleted. While **low-band spectrum** offers great coverage area and wall penetration, there is a big drawback: Peak data speeds will top out around 100 Mbps.
- T-Mobile is the key player when it comes to low-band spectrum. The carrier picked up a **massive amount** of 600 MHz spectrum at a Federal Communications Commission (FCC) auction in 2017 and is using it to quickly build out its nationwide 5G network.
- **Mid-band spectrum** provides faster speeds and lower latency than low-band. It does, however, fail to penetrate buildings as effectively as low-band spectrum. Expect peak speeds upto 1 Gbps on mid-band spectrum.
- **Sprint** has the **majority of unused mid-band spectrum** in the U.S. The carrier is using Massive MIMO to improve penetration and coverage area on the mid-band. Massive MIMO groups multiple antennas onto a single box, and at a single cell tower, to create multiple simultaneous beams to different users. **Sprint** will also use **Beamforming** to bolster 5G service on the mid-band. This sends a single focused signal to each and every user in the cell, and systems using it monitor each user to make sure they have a consistent signal.
- **High-band spectrum** is what delivers the highest performance for 5G, but with major weaknesses. It is often referred to as mmWave. High-band spectrum can offer peak speeds up to 10Gbps and has extremely low latency. The main drawback of high-band is that it has low coverage area and building penetration is poor.

## 6.2.9 Conclusion

- 5G wireless technology is more intelligent technology which will interconnect the entire world without limits.
- It is designed to provide unbelievable and extraordinary data capabilities, unhindered call volumes, and vast data broadcast.
- One would have universal and uninterrupted access the information, communication and entire lives and will change our lifestyle meaningfully.

- Moreover, governments and regulators can use the technology as an opportunity for good governance and can create healthier environments, which definitely, encourage continuing investment in 5G the next generation technology.

### 6.2.10 Advantages

1. Extremely high speed of internet.
2. Faster transmission of images and videos.
3. More reliable connectivity for communications.
4. Better battery life.
5. Low latency, i.e. stop delays.
6. Low battery consumption.
7. Virtual reliability and augmented reliability.
8. Extremely high capacity.
9. Higher resolution and larger bandwidth.
10. Lower cost of infrastructural development.
11. Efficient use of energy in device.
12. Large number of supporting devices.
13. Improved broad band.
14. Improved spectral office.
15. Extremely high reliability.

### 6.2.11 Disadvantages

- When the wireless technology progressed to 4G and 5G wireless technologies, the cells were producing more bandwidth, meaning the coverage radius of each cell was smaller and hence the coverage may drop more often on their 3G wireless network. As the 5G wireless networks gets rolled out, this trend will continue. More call towers will be required to produce this immense bandwidth, because the calls are not able to cover as much space as a 3G or 4G cell. Because of more cells will need to be relieved only so 5G users may not get the coverage area with 5G as wide spread first of with 3G. This is the major disadvantage of *5G wireless* networks.
- The work by 3GPP to define a 4G candidate radio interface technology started in Release 9 with the study phase for LTE-Advanced.

### 6.2.12 Applications

1. Vehicle to vehicle communications could ultimately save thousands of lives.
2. It allows cities and other municipalities to operate more efficiently.
3. Utility companies will be able to easily track usage remotely.
4. Sensors can notify Public Works Departments (PWD), when drains food or streetlights go out.
5. Municipalities will be able to quickly and expensively install surveillance cameras.
6. It will allow technicians with specific skills to control memory from (remote control) in the world.
7. It URLLS ULTRA-Reliable Low Latency Components improves telehealth, remote recovery, and physical therapy vin AR, precision surgery, and remote surgery.
8. Massive machine type communication (mMTC) will display a key role in health care in monitoring patients in hospitals.
9. It is most effectively used for in internet.

## 6.3 COMPARISON OF 4G AND 5G TECHNOLOGIES

Table 6.1

| 4G Technology | 5G Technology |
|---|---|
| 1. It has more capable unified platform. | 1. It has less capable unified platform. |
| 2. It uses spectrum better. | 2. It does not use spectrum better. |
| 3. It is comparatively very fast. | 3. It is comparatively very slow. |
| 4. It has comparatively more capacity. | 4. It has comparatively low capacity. |
| 5. It has very low latency. | 5. It has high latency. |

## 6.4 MOBILE TECHNOLOGY

### 6.4.1 Introduction

- Mobile technology is the technology used for cellular communication. Mobile technology has evolved rapidly over the past few years. Since the start of this millennium, a standard mobile device has gone from being no more than a simple two-way pager to being a mobile phone, GPS navigation device, an embedded web browser and instant messaging client, and a handheld gaming console. Many experts believe that the future of computer technology rests in mobile computing with wireless networking.
- Mobile computing by way of tablet computers are becoming more popular. Tablets are available on the 3G and 4G networks. Mobile technology has different meanings in different aspects, mainly mobile technology in information technology and mobile technology in basketball technology. Mainly based on the wireless technology of wireless devices (including laptops, tablets, mobile phones, etc.) equipment information technology integration.

### 6.4.2 Mobile Communication

- Tesla laid the theoretical foundation for wireless communication in 1890. Marconi, known as the father of radio, first transmitted wireless signals two miles (3.2 km) away in 1894. Mobile technology gave human society Bring great change. The use of mobile technology in government departments can also be traced back to World War I. In recent years, the integration of *mobile communication technology* and *information technology* has made mobile technology the focus of industry attention.
- With the integration of *mobile communication* and *mobile computing technology*, mobile technology has gradually *matured*, and the mobile interaction brought by the *application* and *development* of mobile technology has provided online connection and communication for Ubiquitous *Computing* and *Anytime, anywhere* Liaison and information exchange provide possibilities, provide new opportunities and challenges for mobile work, and promote further changes in social and organizational forms.
- The integration of information technology and communication technology is bringing great changes to our social life. Mobile technology and the Internet have become the main driving forces for the development of information and communication technologies. Through the use of high-coverage mobile communication networks, high-speed wireless networks, and various types of mobile information terminals, the use of mobile technologies has opened up a vast space for mobile interaction and has become a popular and popular way of living and working.
- Due to the attractiveness of mobile interaction and the rapid development of new technologies, mobile information terminals and wireless networks will not be less than the scale and impact of computers and networks in the future. The development of mobile government and mobile commerce has provided new opportunities for further improving the level of city management, improving the level and efficiency of public services, and building a more responsive, efficient, transparent, and responsible government.
- Mobile technology also helps to bridge the digital divide and provide citizens with universal Services, agile service. The integration and development of information and communication technology has spurred the formation of an information society and a knowledge society, and has also led to a user-oriented innovation oriented to a knowledge society, a user-centered society, a stage of social practice, and a feature of mass innovation, joint innovation, and open innovation.

### 6.4.3 Mobile Phone Generations

- In the early 1980s, 1G was introduced as voice-only communication via **"brick phones"**. Later in 1991, the development of 2G introduced Short Message Service (SMS) and Multimedia Messaging Service (MMS) capabilities, allowing picture messages to be sent and received between phones.
- In 1998, 3G was introduced to provide faster data-transmission speeds to support video calling and internet access. 4G was released in 2008 to support more demanding services such as gaming services, HD mobile TV, video conferencing, and 3D TV. 5G technology has been planned for the year 2019.

## 6.5 HIGHER G MOBILE TECHNIQUES

### 6.5.1 Introduction

- The Next Generation Mobile Networks (NGMN) Alliance is a **mobile telecommunications association** of mobile operators, vendors, manufacturers and research institutes. It was founded by **major mobile operators** in 2006 as an open forum to evaluate candidate technologies to develop a common view of solutions for the next evolution of wireless networks. Its objective is to ensure the successful commercial launch of future mobile broadband networks through a roadmap for technology and friendly user trials. Its office is in **Frankfurt, Germany**.
- The NGMN Alliance complements and supports standards organizations by providing a coherent view of what mobile operators require. The alliance's project results have been acknowledged by groups such as the 3rd Generation Partnership Project (3GPP), TeleManagement Forum (TM Forum) and the Institute of Electrical and Electronics Engineers (IEEE).

### 6.5.2 Activities

- The Initial phase of the NGMN Alliance involved working groups on technology, spectrum, intellectual property rights (IPR), ecosystem, and trials, to enable the launch of commercial next generation mobile services in 2010.
- In a white paper first released in March 2006, NGMN summarized a vision for mobile broadband communications and included recommendations as well as requirements. It provided operators´ relative priorities of key system characteristics, system recommendations and detailed requirements for the standards for the next generation of mobile broadband networks, devices and services.
- From July 2007 to February 2008, standards and technologies were evaluated for next generation mobile networks. These were 3GPP Long Term Evolution (LTE) and its System Architecture Evolution (SAE), IEEE 802.16 (products known as WiMax), 802.20, and Ultra Mobile Broadband.
- In June 2008, the NGMN Alliance announced that, *"based on a thorough technology evaluation, the NGMN board concluded that LTE/SAE is the first technology which broadly meets its requirements as defined in the NGMN white paper. The NGMN Alliance, therefore, approves LTE/SAE as its first compliant technology"*. Also in June 2008, the alliance announced it would work with the *Femto Forum* to ensure *femtocells* benefit from the technology.
- In October 2009, the NGMN spectrum working group released *"Next Generation Mobile Networks Spectrum Requirements Update"* (NGMNSRU), containing the status and NGMN views and requirements on frequency bands identified at the ITU WRC-07.
- Since next generation devices, networks and services need to be synchronized for a successful launch, NGMN in February 2009 released a *'whitepaper'* which provided generic definitions for next generation (data only) devices to ensure that devices were available at the time, when first networks were launched in 2010.
- After the launch of the first LTE networks in 2010, the alliance then addressed challenges of network deployment, operations, and interworking, while focusing on LTE and its evolved packet core, as defined by the SAE. In September 2010, NGMN published recommendations on operational aspects of next generation networks.

- In 2014, the NGMN Board decided to focus future NGMN activities on defining the end-to-end requirements for 5G. A global team has developed the NGMN 5G *'Whitepaper'* published March 2015 delivering consolidated operator requirements that will support the standardisation and development of 5G. NGMN encourages the industry to have 5G solutions available by 2020.
- In 2015, NGMN launched a 5G-focused work-programme that will build on and further evolve the *Whitepaper guidelines*. The main 5G NGMN work-items for 2015 are; the development of technical 5G requirements and architectural design principles, the analysis of potential 5G solutions and, the assessment of future use-cases and business models. Furthermore, the NGMN project teams will address the areas TDR and Spectrum from a 5G perspective. In September 2015, the NGMN published a Question and Answer about 5G.

### 6.5.3 Next Generation Super Phone

- Looking at the history of mobile phone development, the mobile phone revolution runs on a 12 years cycle, evolving in a way synonymous with changing relationship between man and device. Motorola's first generation (1G) feature phone appeared in 1995 and instantly became the last for an entirely new form of communication. Making calls or texting with a mobile phone changed the way people communicate and brought us closer to one another. At that time, the most distinctive feature of the mobile the big external antenna.
- Twelve years later in 2007, come the arrival of the iphone followed by the birth of an android smartphone in 2008, making a significant leap from the feature phone to the smartphone. These were the first generation (1G) of mobile phones to run one operating system. And so from a loop for simple communications, mobile phones evolved into assistant providing entertainments, shopping, social networking and more, all of which have now become a necessary in a in our daily lives.
- In the next 12 years cycle, we can boldly predict that the relationship between mobile phones and human beings will become even closer, evolving towards a synthesis if man with his world. The next generation of mobile phones will be the 'SUPER PHONES' emerging in approximately 2020 to 2021, and unveiling a rewera, connecting physical and virtual realities.
- 'SUPER PHONE' will be living organization, like our families and friends recording our behaviors and habits, and understanding our likes and references. This device will be able to understand and interact with human and the environment around it making our smart and simple life.
- 'SUPER PHONE' will connect the world with to the world of innovatine technologies through virtual reality. Though 'sensor systems' technology and 3D-Rays, the device will be able to capture human behavior. This device will "Empower" a normal human in following ways :

  1. Acquire the Six Human Senses.
  2. Creating the "Super human" via the (IoT).
  3. A symbiotic partnership of man and device.
  4. SUPER PHONE and human interaction.
  5. Texting by Thinking.

### 6.5.4 Benefits of Super Phone

1. **Health Management:** Human will be able to manage his/her health parameters like blood sugar and blood pressure level etc. and can take proactive action to remediate the health issues.
2. **Air Pollution Monitoring:** Through this device, human will be able to check the surrounding 'Air Quality' and take the next steps.
3. **Geographical Positioning:** Using this feature of device, human will be able to reduce his dependency on maps.
4. **Man-Object Interaction:** Using this device, human and object interaction will be taken to next level where he/she can control any object using this device.
5. **Reminder Notification:** This device will also send reminder notifications of up-coming travel bookings by analyzing human's flight/hotel bookings. This device will also notify regarding any day-to-day daily activity by analyzing human's daily schedule.

6. **Virtual Communication:** This device will allow humans to have face to face virtual communication with other humans so as to attend any virtual training or have a virtual doctor visit.
7. **Virtual Store:** Using this device, humans will be able to dry cloths and will be able to buy/exchange any cloth virtually.

## 6.6 LTE-ADVANCED SYSTEMS

### 6.6.1 Introduction

- LTE Advanced is a mobile communication standard and a major enhancement of the Long Term Evolution (LTE) standard. It was formally submitted as a candidate 4G to ITU-T in late 2009 as meeting the requirements of the IMT-Advanced standard, and was standardized by the Third Generation Partnership Project (3GPP) in March 2011 as 3GPP Release 10.
- The LTE format was first proposed by NTT DoCoMo of Japan and has been adopted as the international standard. LTE standardization has matured to a state, where changes in the specification are limited to corrections and bug fixes. The first commercial services were launched in Sweden and Norway in December 2009.
- More LTE networks were deployed globally during 2010 as a natural evolution of several 2G and 3G systems, including Global system for mobile communications (GSM) and Universal Mobile Telecommunications System (UMTS) in the 3GPP family as well as CDMA2000 in the 3GPP2 family. In the field of mobile communication services, the 4G mobile services are the advanced version of the 3G mobile communication services.
- The 4G mobile communication services are expected to provide broadband large capacity, high-speed data transmission, providing users with high quality colour video images, 3D graphic animation games, and audio services in 5.1 channels.
- We have been researching the vision of 4G mobile communication systems services and architectures. We also have been developing the terminal protocol technology for high capacity, high speed packet services, public software platform technology that enables downloading application programs, multimode radio access platform technology and high quality media coding technology over mobile network
  1. Support interactive multimedia services like teleconferencing wireless Internet etc.
  2. Wider bandwidths, higher bit rates.
  3. Low cost.
  4. Scalability of mobile networks.
- By 2009, it had become clear that at some point, 3G wireless networks would be overwhelmed by the growth of bandwidth-intensive applications like streaming media. Consequently, the industry began looking to data optimized fourth generation (4G) technologies, with the promise of speed improvement upto 10 fold over existing 3G technologies. Out of the main ways in which 4G differed technologically from 3G in its elimination *of circuit switching*, instead employing on all IP networks.
- Thus, 4G ushered in a treatment of voice calls just like any other type of streaming audio media, utilizing packet switching over internet, LAN or WAN networks via VoIP.

### 6.6.2 4G LTE Wireless Network

- LTE is an abbreviation for **L**ong **T**erm **E**volution. LTE is a 4G wireless communication standard developed by the 3$^{rd}$ Generation Partnership Project (3GPP) that is designed to provide upto 10 times the speeds of 3G wireless networks for mobile devices such as smart phones, tablets, notebooks, netbooks and wireless hotspots.
- 4G technologies are designed to provide IP based voice, data and multimedia streaming at speeds of at least 100 M bit per second and upto as fast as 1 GB per second. 4G LTE is one of the several competing 4G standards along with Ultra Mobile Broadband (UMB) and Wi-Max (IEEE 802.16).

- The leading cellular providers have started to deploy 4G technologies, with Verizon and AT and T launching 4G LTE networks and sprint utilizing its new 4G Wi-Max network. In terms of mobile devices, many newer Android based smartphones are 4G LTE capable and both the iphone 5 and the ipad 3 are expected to have built-in 4G LTE capabilities when released in the second half of 2012.

### 6.6.3 Need

- The need for the 4G wireless technology is far much more broadband capacity, high-speed, data transmission, high quality colour video images, 3D graphic, animation games and low cost requirements to overcome the disadvantages or lack of functionality provided by 3G wireless technologies.

### 6.6.4 Definition

- 4G is not one defined wireless technology or standard, but rather a collection of technologies at creating fully packet switching networks optimized for data.
- 4G wireless networks are projected to provide speed of 100 Mbps while moving and 1 Gbps while stationary.

### 6.6.5 Concept of 4G

- At present the download speed for mode data is limited to 9.6 kbit/sec, which is about 6 times slower than ISDN (Integrated Services Digital Network) fixed line connection. Recently, with 5041 handsets the download data rate was increased from 3 fold to 28.8 kbps. However, in actual use the data rate are usually slower especially in crowded areas or when the network is 'congested'.
- For third generation mobile (3G FDMA) data rates are 834 kbps (download) maximum, typically around 200 kbps, and 64 kbps upload since spring 2001. Fourth generation (4G) mobile communications will have higher data transmission rates than third generation (3G). 4G mobile data transmission rates are planned to be upto 20 Mbps.
- **LTE Advanced is a mobile communication standard and a major enhancement of the Long Term Evolution (LTE) standard.**

### 6.6.6 Technical Considerations

1. Continual improvement to the LTE radio technology and architecture.
2. Scenarios and performance requirements for working with legacy radio technologies.
3. Backward compatibility of LTE-Advanced with LTE. An LTE terminal should be able to work in an LTE-Advanced network and vice versa. Any exceptions will be considered by 3GPP.
4. Consideration of recent World Radio Communication Conference (WRC-07) decisions regarding frequency bands to ensure that LTE-Advanced accommodates the geographically available spectrum for channels above 20 MHz. Also, specifications must recognize those parts of the world in which wideband channels are not available.

### 6.6.7 Deployment

- The deployment of LTE-Advanced in progress in various LTE networks.
- In August 2019, the Global mobile Suppliers Association (GSA) reported that there were 304 commercially launched LTE-Advanced networks in 134 countries. Overall, 335 operators are investing in LTE-Advanced (in the form of tests, trials, deployments or commercial service provision) in 141 countries.

### 6.6.8 Advantages

1. Faster and more reliable.
2. Speed upto 100 Mbps.
3. Lower cost than previous generations.
4. Multi-standard wireless system.
5. Bluetooth, wired, wireless.
6. Ad-hoc networking.
7. OFDM used instead of CDMA.
8. IPV6 core.
9. Potentially IEEE standard 86.02- 11 n.
10. Most information is proprietary.

### 6.6.9 Main Aspects

- The Fourth Generation (4G) is ready for implementation before the end of 2013. The main aspects of 4G are as under :
  1. With the case of 4G in addition to that of the services of 3G have some additional features such as multi-media newspaper.
  2. You can also watch TV programs with the clarity that of an ordinary TV.
  3. 3.In addition, we can send data much faster than that of the previous generations.
  4. A 4G cellular system must have target peak, data rates of upto approximately 100 Mbps for high mobility such as mobile access and upto approximately 1 Gbps. for low mobility such as per nomadic/local wireless access according to the ITU requirements.
  5. Scalable bandwidths upto at least 40 MHz should be provided.
  6. A 4G system is expected to provide a comprehensive and secure all the IP based solution, where facilities such as IP telephony, ultra broadband Internet access, gaming services and HDTV streamed multimedia may be provided to users.
  7. Although legacy systems are in place to adopt existing users, the infrastructure for 4G will be only packet based (all IP).
  8. Some proposals suggest having an open Internet platform technologies considered to be early 4G include Flash OFDM, the 802.16e mobile version of WiMax (also known as WiBro in South Korea, and HC-SDMA.

**Practice Questions**

1. What is 4G technology ?
2. What are key features of 4G technology ?
3. State advantages and disadvantages of 4G and 5G.
4. What is 5G technology ?
5. Why do you need 5G ?
6. State basic concepts of 5G technology.
7. Explain working principle of 5G technology.
8. List important applications of 5G technology.
9. State advantages of 5G technology.
10. Compare 4G and 5G technologies.
11. What is higher (or next) generation technique ?
12. Describe next generation superphone.
13. List benefits of superphone.
14. What is LTE advanced (4G) wireless network ?
15. List technical considerations of 4G LTE wireless network.
16. State advantages of 4G LTE wireless network.
17. What are basic requirements of 5G technology ?
18. State key features of 5G technology.

✍✍✍

www.ingramcontent.com/pod-product-compliance
Lightning Source LLC
LaVergne TN
LVHW080455160826
845677LV00006B/1364